An Introduction to U.S. Federal Public Lands

This book provides a comprehensive and engaging introduction to the U.S. federal land system, explaining how public lands are used and managed.

The federal government owns more than a quarter of all American land, encompassing national parks, forests, wildlife preserves, and monuments. These federal lands are an essential feature of America's politics, culture, and economy, and their management is at the core of U.S. environmental policy. This book describes the law and policy of land management to enable readers to engage with the public lands. It begins with an ecological and historical overview before detailing the categories of public lands as well as the various agencies that oversee them, including the National Park Service, the Fish and Wildlife Service, the Forest Service, and the Bureau of Land Management. It explains the basic legal authority for land management and discusses the current priorities codified in American law. The final chapters discuss key issues, questions, and challenges facing federal lands, such as climate change, wildfire, biodiversity, recreation, and mining. Each chapter concludes with study questions and recommendations for further reading.

This book will be essential reading for students, scholars, and citizens interested in the federal lands, natural resource management, human geography, public policy, and environmental law. It will be of great use to those who have or envision a career in land and resource management, particularly in the United States.

Bruce R. Huber is Professor of Law at the University of Notre Dame, USA. He is the author of many law review articles and book chapters and the co-editor of *Energy Law and Economics* (2018) and *Environmental Law and Economics* (2017).

James R. Skillen is Professor of Environmental Studies at Calvin University, USA, and Director of the Calvin Ecosystem Preserve and Native Gardens. He is the author of *This Land Is My Land* (2020), *Federal Ecosystem Management* (2015), and *The Nation's Largest Landlord: The Bureau of Land Management in the American West* (2009).

Earthscan Studies in Natural Resource Management

Governing Natural Resources for Sustainable Peace in Africa
Environmental Justice and Conflict Resolution
Edited by Obasesam Okoi and Victoria R Nalule

Capacity-Building and the Water-Energy-Food Nexus
Rethinking Integration in the Asia-Pacific
Maureen Papas

Governing the Palm Oil Industry
Perspectives from Southeast Asia and Latin America
Edited by Patrick O'Reilly and Helena Varkkey

Researching Institutions in Natural Resource Governance
Methods and Frameworks
Edited by Fiona Nunan

Sustainable Management of Cordyceps
Supply Chains and Resource Management Policies
Edited by Jiping Sheng and Ksenia Gerasimova

Reimagining Resources and Community Development
Lessons from Newfoundland and Labrador
Monica Gruezmacher, Kelly Vodden, Brennan Lowery, Amy Hudson and Kristof Van Assche

An Introduction to Land Tenure in Sub-Saharan Africa
Dualism, Reform and the Roots of Conflict
John W. Bruce

Natural Resource Cooperatives in Africa
Sustainable Management and Governance
Edited by Andrew Emmanuel Okem, Terence Muzorewa and Ibrahima Barry

An Introduction to U.S. Federal Public Lands
Bruce R. Huber and James R. Skillen

For more information about this series, please visit: www.routledge.com/Earthscan-Studies-in-Natural-Resource-Management/book-series/ECNRM

An Introduction to U.S. Federal Public Lands

Bruce R. Huber and James R. Skillen

Routledge
Taylor & Francis Group
LONDON AND NEW YORK

Designed cover image: Getty Images

First published 2026
by Routledge
4 Park Square, Milton Park, Abingdon, Oxon OX14 4RN

and by Routledge
605 Third Avenue, New York, NY 10158

Routledge is an imprint of the Taylor & Francis Group, an informa business

British Library Cataloguing-in-Publication Data
A catalogue record for this book is available from the British Library

Library of Congress Cataloging-in-Publication Data
A catalog record has been requested for this book

ISBN: 978-1-032-49185-1 (hbk)
ISBN: 978-1-032-47395-6 (pbk)
ISBN: 978-1-003-39251-4 (ebk)

DOI: 10.4324/9781003392514

Typeset in Times New Roman
by Taylor & Francis Books

To those who have devoted their careers to the public lands

Contents

Figures

Tables

Boxes

Preface

We have studied federal public lands for the last two decades, and like so many others, we have enjoyed a host of benefits from them. We have come to know and respect many of the diligent federal employees who have devoted their careers to national parks, forests, and wildlife refuges. As scholars, we have traced the conflicting public demands on public lands and the evolution of public land law and policy. We have witnessed profound shifts in public land use, from dramatic increases in outdoor recreation to expanded leasing for solar installations, oil and gas production, and lithium mining.

For most of our academic careers, federal public land law and policy has been fairly stable. Congress has been largely content to preserve the legal foundations laid in the late nineteenth and early twentieth centuries, and has not made major legislative changes to public land law since the 1970s. Indeed, over the last 50 years, Congress has generally allowed the courts and the president to guide the evolution of public land policy and management. From President Reagan to President Obama, public land agencies were held in a rough equilibrium between the ideologies of two relatively stable political coalitions. Republicans typically argued for greater access to (and resource production from) public lands, and they worked to roll back environmental regulations they found in their way. Democrats argued for greater environmental protection, favoring biodiversity protection and nonconsumptive outdoor recreation to logging and grazing. Whatever their rhetoric, both parties supported fossil fuel development on public lands.

We began writing this book when President Biden was in office and Congress was divided. It looked like the status quo in public land politics and management would hold. But over the course of writing, we have seen shifts of greater magnitude than at any point in our careers. In some cases, of course, it is difficult to see through the rhetorical bluster to measure substantive change. In many other cases, the changes are obvious, concrete, and dramatic. Much of the dynamism stems from a restructuring of the Republican Party's ideological coalition, evident in the Trump Administration and in Congress. Reagan-era commitments to free trade, global intervention, and fiscal responsibility have given way to populist protectionism and an austere, albeit ambivalent, fiscal ideology. The effects on public lands are both direct and indirect. The Trump Administration made deep cuts in public land agency staffing; the permanent staff of the Forest Service, for example, was reduced by more than 25%. The Administration's expansive use of tariffs, while not directly targeting public lands, created significant uncertainty for natural resource markets. To top it off, recent decisions of the U.S. Supreme Court have enhanced presidential control over the federal bureaucracy. We can expect that future presidents will have an easier time making sizeable changes to federal land policy.

These shifts create a challenging context for the release of an introduction to federal public lands. Permanent federal land ownership is an old institution, dating back at least to Yellowstone National Park's creation in 1872. The four main federal land agencies were established between 1905 and 1946. Federal public lands, and public land agencies, are embedded in a complex political economy with enormous inertia. Too many state governments, industries, and individual users depend on public lands to imagine them being sold en masse. Gridlock in Congress is so profound that it is hard to imagine a legislative overhaul in the public lands arena. But it is still difficult to speak confidently about the direction that public land management will take in the near future.

Whatever the future holds, we trust that most of this book's content will remain relevant. The history of federal public lands isn't changing, and we are confident in our basic assessment of the public land agencies and the issues they face. At the same time, while writing we had to revise the text often to reflect the considerable uncertainties in public land management. There is no reason to assume that dynamism will give way to stasis. While issues like biodiversity protection and climate change will remain paramount for federal land agencies, federal policy in these areas is likely to remain unstable.

In sum, the current dynamics in federal public land law and policy have forced us to do what we know best as scholars, which is interpreting the history of public lands and their present management. But these dynamics have caused us to shy away from predictions that we might have made with modest confidence 10 years ago.

More importantly, dynamism does not diminish the significance of federal public lands culturally, ecologically, economically, or politically. The federal government owns about 28% of all land in the United States, most of it in the western states, all but guaranteeing that the public lands will remain a defining feature of the West's political economy. Indeed, federal public lands are an economic engine in the West, and they shape lifestyles in the West in dramatic ways. And many public lands, particularly national parks, remain a point of pride for countless Americans.

Throughout this book, we have done our best to emphasize the rich history, contemporary significance, and durability of federal public lands while simultaneously identifying the uncertainties ahead. Our efforts have been guided most significantly by one audience: undergraduate and graduate students curious about the public land tradition. We are confident that this book offers students a solid introduction to the basic history, legal structure, and management challenges of federal public lands. We are hopeful that we've also given students a clear sense of the dynamic context in which public land policy is made. We have other audiences in mind, too, including readers who want a basic primer on federal public lands with sufficient depth to help them navigate their own use and enjoyment of the federal estate. For them, this book will go a long way toward explaining the many facets and idiosyncrasies of the national parks, forests, and wildlife refuges. Most of all, we hope this book will foster in readers an appreciation for the difficult task of public land management—a task carried out each day by public servants, who bear the weight of a tangled and tainted legacy while stewarding the land for future generations. It is to them that we dedicate this book.

Acknowledgments

We wish to thank the many members of the public lands community who have shared their wisdom and experience with us over the years. This includes federal employees, scholars of law and history, scientists, and countless others committed to the land's protection. We also owe an enormous debt of gratitude to Addison Arendt for lending us his remarkable cartographic talent. This book reflects the excellent work of a handful of top-notch research assistants at Calvin University and Notre Dame Law School: Summer Abdelbarry, Hayden Burnight, Katelyn Cothren, Josh Hill, Silas Kok, Aleah Schrock, and Gloria Wertjes. We thank Courtney Schultz, our editorial team, and several anonymous reviewers for commenting on portions of the manuscript and sparing us many errors and oversights. Finally, thank you to our students, for their curiosity and enthusiasm; to our colleagues, for their intellectual and professional support; our friends, for their encouragement and companionship; and our families, for their love and patience.

1 Introduction

The real estate market in California is absurd. Modest homes around Los Angeles, San Diego, and the San Francisco Bay regularly sell for well over one million dollars. McMansions that would fetch $400,000 in midsize Midwestern cities draw bids for ten or twenty times that amount in urban California. Speculative bubbles come and go, but real estate in the Golden State has been reliably minting millionaires for at least half a century. The staggering concentration of wealth represented by California land is a social phenomenon that will be studied for decades to come.

Yet there is another social phenomenon, also related to western land, that may in the long run be of greater human significance. Just outside the cities of California are vast areas of public land—areas that are every bit as valuable economically but that are open to the public and titled in the U.S. federal government. Barely one mile north of San Francisco, across the historic Golden Gate, is an undeveloped federal recreation area well over twice the size of the *entire city*. Here one can hike alone amidst windswept grasses while looking down on rocky Pacific beaches. Ridgelines extend up the coastal hills, uninterrupted by homes or city streets. Similarly, just north of Los Angeles, the rugged San Gabriel National Monument offers a vast mountainous retreat unmarred by SoCal sprawl. Flying into LAX, the juxtaposition between the verdant San Gabriels and the gargantuan city below is jarring. The land in these undeveloped areas, perched adjacent to some of the world's most glamorous cities, is literally worth trillions of dollars—yet We The People have decided to keep it public, open, unbuilt, and wild.

A bit farther beyond the cities' orbit, the scale of public land ownership becomes difficult to grasp. Running north and west from the Los Angeles basin, the Los Padres National Forest covers nearly 3,000 square miles, an area over half the size of Connecticut. And the Los Padres is small compared to the federal landholdings in northern and eastern California, along the Sierra Nevada range, into Death Valley and southward through the Mojave Desert and the Imperial Valley. In these portions of the state, private land ownership is the exception. Instead of sprawling estates or mountain retreats, there are Forest Service campgrounds on unpaved logging roads, a handful of world-class national parks, massive military training grounds, and desolate desert landscapes. California is the most populous state in the U.S., yet almost *half* its land is owned by the public—that is, by the federal government.

In other western states, federal ownership is even more extensive. Public lands cover roughly half of Arizona, Utah, Idaho, and Oregon, and a much larger share of Nevada and Alaska. From the Rockies to the Pacific, federal land ownership is an unavoidable fact of life. Federal agencies that are barely familiar to easterners—the Bureau of Land Management, the Fish and Wildlife Service, the Forest Service—are omnipresent neighbors in much of the American West.

DOI: 10.4324/9781003392514-1

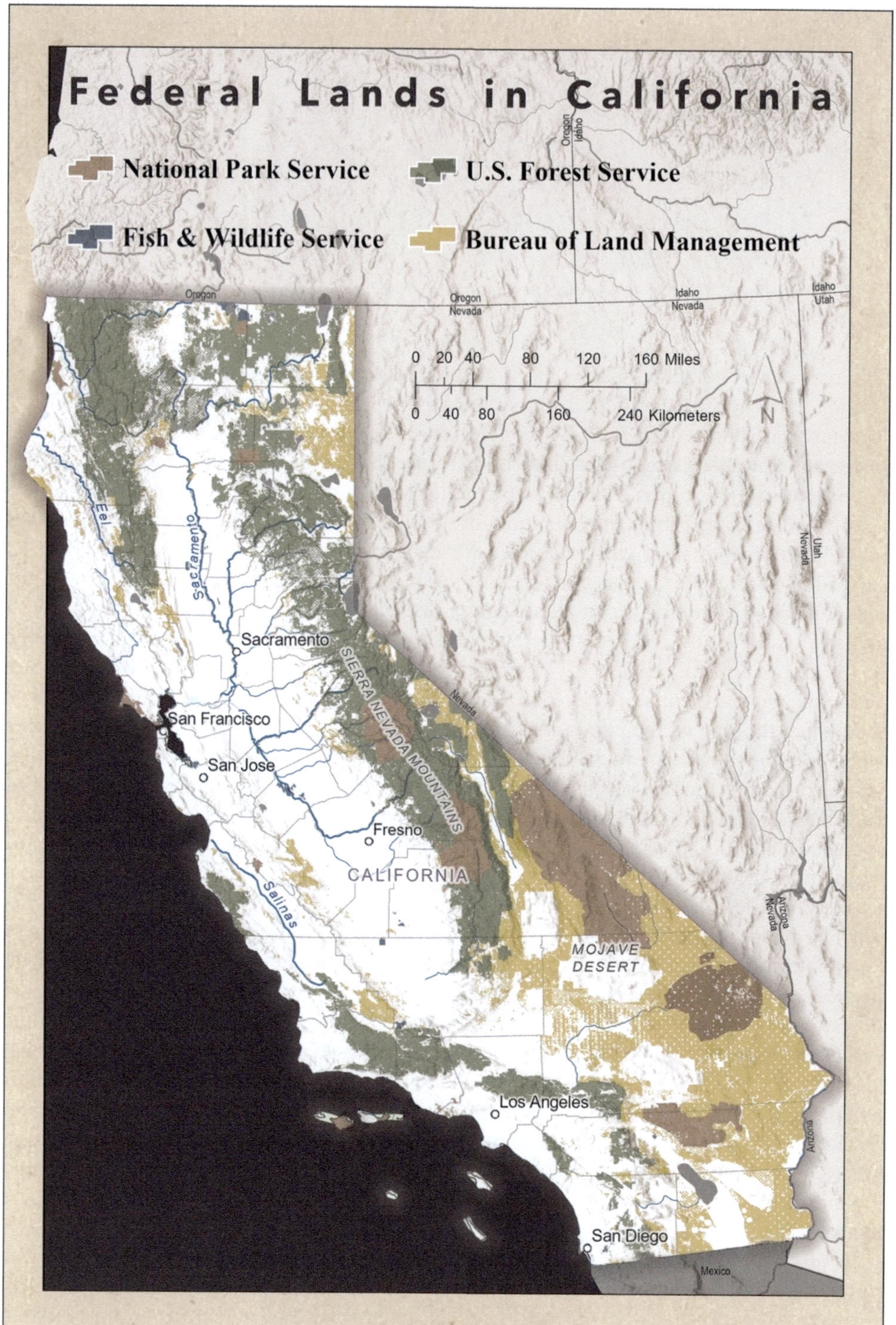

Figure 1.1 Federal lands in California. Map by Addison Arendt.

This book is about the U.S. public lands—how they came about, how they are managed, and what they mean. They include national parks and forests, wildlife preserves and wilderness, and much else. Public lands are used in infinite ways. They offer places to camp, minerals to mine, oil to drill, rivers to raft, game to hunt, and landscapes to survey. They represent a vast natural heritage, a heritage owned by all Americans in common through their central government.

In this introductory chapter, we will provide a very brief overview of the federal public lands. The chapter first offers some context to help you understand why public land is important. Next, it explains why federal ownership is minimal in some regions in the country yet extensive in others. Finally, the chapter will describe how the rest of the book is organized.

The Public Lands Are a Choice

Public lands represent a policy choice—an *ongoing* policy choice whose dimensions and implications we seldom consider. Indeed, widespread public land ownership is one of the most underappreciated and ironic aspects of U.S. history, policy, and government. American political culture sacralizes private property and regards federal power with suspicion, yet that same culture created a system whereby federal agencies manage fully half of the American West. Any political fact so discordant with its context cannot come about without a long list of causal forces and an equally long list of sustaining ones. To interpret the public lands is to inquire into and describe those forces and their interaction, and that is one of the tasks of this book.

Not everyone endorses federal land ownership. The public lands have enemies. Some people object to specific decisions about land, such as a decision to restrict hunting in a nearby forest. Others oppose management by far-off federal bureaucrats and would prefer public lands to be under local control. Still others bemoan the fact that federal lands don't produce property tax revenue. For these reasons and more, many voters wish that the federal government would give its lands to state governments. "When an unelected and unaccountable bureaucracy owns and manages more than half the land in your state, that is a recipe for disaster."[1] So said Senator Mike Lee of Utah, a state that has historically regarded the federal public lands with some ambivalence. On one hand, Utah's national parks attract countless tourists. On the other hand, much of Utah's land is managed by people outside the state. Residents of Alaska, too, are notoriously suspicious of their federal landlords; employees of the National Park Service in Alaska report hostile glares and unfriendly encounters with longtime Alaskans at grocery stores and gas stations. In Idaho, Montana, Wyoming, and elsewhere across the West, federal land ownership is a recurring topic of concern during election cycles. Strong opinions abound.

The Public Lands Mean Many Things

But the mere fact of public land ownership is only half the story. Regardless of ownership, the public lands themselves are awash in significance—to those who use them, to those who live near them, to those who visit them or dream of doing so. Like the proverbial onion, the public lands have many layers; a thousand narratives could be assembled and still fail to capture the myriad angles that give the public lands their meaning. Just as a beloved relative's home occupies an outsized portion of one's childhood imagination, so also does public land assume a prominent place in the minds and hearts of those who spend time there—but that place will be different for the forester than it is for the Native American, and different for the rancher than it is for the backpacker. The public lands belong to all and represent something different to each one.

Here are just a few of the disparate frames through which one might view the public lands today.

- *Scenic beauty.* For many, the most salient fact about the U.S. public lands remains their unparalleled beauty. To be sure, millions of federal acres are flat and unmemorable. But taken as a whole, the public lands are a spectacular gallery of gems, as demonstrated by the gawking hordes of visitors they attract. Certain iconic vistas are so beloved that they exert a trump-card power in public land policy: the mere mention of, say, uranium mining in the vicinity of the Grand Canyon elicits a fierce and protective reaction from the American public. Even in desolate places far off the beaten path, scenic viewsheds command an immediate constituency. Especially in the age of social media, a single photograph can trigger a wave of preservationism. With such beauty comes traffic, of course, and Instagram may have imperiled as many sites through overcrowding as it has helped to protect. But scenery is undeniably one of the core facts to which public land management must attend.
- *Fire.* In recent years, the public lands have been ravaged by some of the fiercest wildfires the West has ever witnessed. Fires of this severity are often the product of both land mismanagement and extraordinary drought, which is itself a likely consequence of climate change. In fact, there is no more potent symbol of climate concern than western wildfire. The destructiveness of fire is measured also by its economic costs, which have risen as more and more homebuilders encroach upon fuel-rich forestland. Firefighting costs now dominate forest management expenditures, and the wildfire season rivals the hurricane season in its media coverage. Now more than ever before, public land management *is* wildfire management.
- *Mineral resources.* One prevalent political narrative about public lands has to do with the oil, gas, and minerals that lie beneath them. Since the first days of the California Gold Rush, which took place on federal land, Americans have fantasized about hidden lodes of fabulous mineral wealth, buried deep in the continent's interior. In the modern era, those fantasies are an everyday reality: billions upon billions of dollars' worth of gold, silver, copper, uranium, natural gas, petroleum, timber, and much else is removed from federal land by private enterprises. Some protest the environmental consequences or fear that these resources are being underpriced, but others credit the federal government's extraordinary open-handedness with invigorating a world-leading natural resource economy.
- *Overcrowding.* Tourism and recreation dominate the public's relationship to the public lands. Vacations to Yellowstone, the Grand Canyon, or the Smoky Mountains are summertime staples, and increasingly ones in which foreign visitors also partake. But with exploding interest comes exploding crowds. The growing pains are most acute in the big-name national parks, where many attractions now require reservations, timed-entry passes, permits, or other annoyances. Neighboring "gateway communities" reap economic benefits but suffer predictable consequences: impossible parking, seasonal housing shortages, and a need for additional law

enforcement. Even outside national parks, social media can transform a little-known treasure into a must-see destination in a matter of days. Land managers are in a constant race against the clock, trying to channel crowds before they cause damage to land. The footprint of tourism can be found even in previously desolate locales, leaving many to wonder whether any genuine wilderness can long remain.

What Lands Are Public Lands?

The North American landscape is remarkable in its ecological diversity. Northern Alaska extends above the Arctic Circle; it receives no sunlight in the heart of winter. Temperatures remain below freezing for well over half of the year, and it is surprisingly dry, with just 5 inches of annual precipitation. Southern Florida, by contrast, rarely sees frost. The region never has less than eleven hours of sunlight and receives *60* inches of annual precipitation. And one need not travel thousands of miles to find striking differences in climate, flora, and fauna. The summit of Mount Whitney, located on the eastern edge of Sequoia National Park, is just 80 miles west of Death Valley National Park in California. The aptly named Death Valley contains the lowest point in the United States (some 282 feet below sea level), has registered the highest ambient air temperature ever recorded on Earth (134° F), and is one of the driest locations in North America (less than 2 inches of annual rainfall). Yet Mount Whitney, at 14,505 feet, is the *highest* point in the Lower 48 states. Temperatures there rarely reach 80° F, and the mountain receives over 40 inches of precipitation, much of it in the form of snow. The proximity between these extremes inspired the Badwater Ultramarathon, which takes place every July and challenges contestants to run the 135 miles from Death Valley to Mt. Whitney. (Competitors are disqualified if they need an emergency IV, so stay hydrated!)

Given this extraordinary diversity, which lands became today's *public* lands? As we'll see throughout this book, the answer has much to do with patterns of human settlement. Native peoples were able to flourish across the continent, thriving in wide-ranging environments. They did so by adapting their lifestyle to the distinct rhythms of particular landscapes and altering those landscapes in important ways, resulting in diverse cultures and technologies among the various people groups. But as European Americans moved west from the Atlantic coast in the nineteenth century, they displaced Native Americans and reshaped the landscape with a different, and more homogeneous, cultural imagination and with quite different technology.

As you might expect, these settlers found some areas more useful, desirable, or habitable than others. Farmers plowed the rich prairie lands of Iowa but rushed past the hot deserts of Nevada. Timber companies rapidly developed the Great Lakes forests, where they could easily bring lumber to market, but they long ignored forests in the Rocky Mountains. And miners obviously could only mine where they found valuable mineral deposits. The distribution of federal public lands in the United States is, to a considerable extent, a reflection of this inconsistent settlement and development. As we'll describe in more detail in subsequent chapters, most of today's public lands are those that European Americans did not settle or develop in the nineteenth century. It should come as no surprise, then, that the federal government owns more desert than prairie, more tundra than midwestern forests. Figures 1.3 and 1.4, later in the chapter, show this lopsided ownership pattern.

The American Landscape

To appreciate the pattern of settlement in the Lower 48 states, it's helpful to imagine how European Americans in the nineteenth century encountered the North American landscape. The map in Figure 1.2 depicts one widely-used representation of the continent's ecological regions (or "ecoregions"). These regions are defined by a common climate and by common plant and animal communities. Obviously, there is significant variation within each ecoregion, but this broad system of classification will help us see how human settlement was affected by large-scale differences in climate and ecology. We will adopt the names and numbers of the regions shown on the map because this labeling system is widely used in scientific literature. Let's work our way through the ecoregions of the United States, moving from east to west—and then north to Alaska—to match the path of European American settlement.

Ecoregion 8 (Eastern Temperate Forest)

As you can see, nearly the entire eastern half of the present United States falls within one large ecoregion labeled "Eastern Temperate Forest." This ecoregion stretches from the Great Lakes to the Gulf of Mexico and from the Atlantic Coast to eastern Texas. Topographically, the region is relatively moderate, although it includes the Appalachian Mountains, which rise to more than 6,500 feet. Geologically, it contains rich mineral resources, including coal and iron, as well as productive soil, owing to the sedimentary material on the coastal plains and undulating hills of what is now Ohio and Indiana. The area's climate is humid but temperate; both temperature and precipitation increase

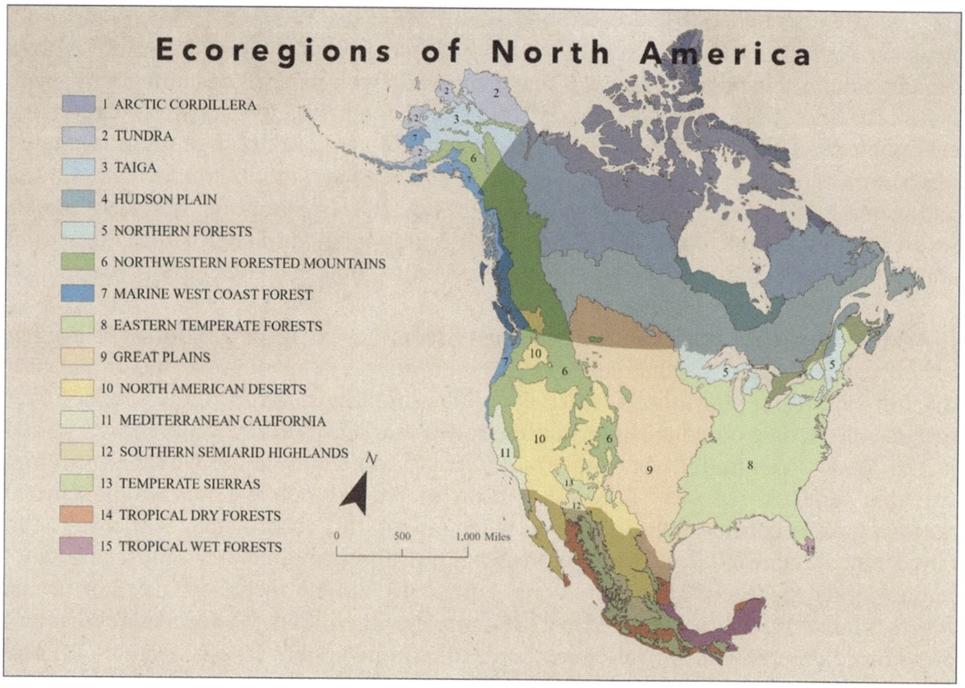

Figure 1.2 Ecoregions of North America. Map by Addison Arendt.[2]

within the region from north to south. Unsurprisingly, the growing season is considerably shorter in northern New Hampshire than it is in Florida. With excellent shelter and rich sources of food, eastern forests supported and continue to support a remarkable diversity of mammals, birds, amphibians, and reptiles, particularly in the southern half of the region. The region's eastern and western edges serve as two of the most significant North American flyways for migratory birds: the Atlantic Flyway and the Mississippi Flyway.

This diverse ecoregion, in other words, held all of the essential natural resources for European American settlement. So as Native Americans were dispossessed and as transportation improved—by way of ships on the Great Lakes and Mississippi River, and an extensive rail system in the second half of the nineteenth century—nearly all of the land within this enormous region came to be privately owned. Today the federal government owns just 4% of the ecoregion, almost all of which it purchased back from private or state ownership. There are very few large, ecologically intact areas.

Ecoregion 5 (Northern Forest)

Similarly, the federal government owns only a small fraction of the acreage within the "Northern Forest" ecoregion. This region extends broadly across Canada, dipping into the United States only in the northern Appalachian Mountains and the northern Great Lakes states. Northern forest in the United States is largely a transition zone between the boreal conifer forests of Canada and the deciduous forests to the south. It is a patchwork of spruce, pine, fir, and broad-leafed trees, such as maple, beech, birch, and oak. The climate, though moderated by maritime air masses off the Atlantic Ocean, is much colder than the eastern temperate forest, with cool, moderately humid summers and harsh winters. The forests are rich with wildlife, such as deer, moose, and black bears, as well as a wide range of birds and small mammals. Even today, the region remains largely rural and forested. Significant portions of the northern forest were owned by the original colonies and never entered federal ownership. Whatever land the federal government did acquire was quickly sold for logging. As a result, the federal government today owns just 11% of the ecoregion's land. The most notable protected land in this ecoregion may be the 6-million-acre Adirondack State Park in New York.

Ecoregion 9 (Great Plains)

The Great Plains ecoregion stretches from Canada to the southern end of Texas. Its western edge is the Rocky Mountains. Erosion carried sediments from west to east across the plains, and glaciers further eroded the northern portions of the region, carving distinct landforms in what is now South Dakota and Wyoming. There you can find the underlying igneous and metamorphic rock that has been exposed in places like the Black Hills and Devils Tower. Glacial soils and moraines dominate the northern plains, which are dotted with lakes.

Ecologically, the Great Plains were ideal for European American agriculture, with excellent soils and adequate rainfall. The soils supported vast short grass prairies to the west and tall grass prairies to the east. The absence of forest—which in this region is limited primarily to river bottoms—facilitated agriculture. The prairies supported vast herds of bison, elk, pronghorn, and deer, which in turn supported a variety of predators. And the northern lakes provided critical roosting and resting places for flocks of migratory birds.

As railroads linked the Great Plains to Chicago and other markets, the federal government used various tactics to remove and relocate Native Americans. Having thus cleared the title, the government transferred virtually all the region's land into state and private ownership. As a result, the federal government today owns just 3% of the region, and very little of the native flora and fauna remain. Tall grass prairie can only be found in a few isolated reserves, and bison, pronghorn, and elk were largely extirpated to make way for cropland and domestic livestock.

Ecoregion 6 (Northwestern Forested Mountains)

Ecoregion 6, labeled "Northwestern Forested Mountains," dominates the Rocky Mountains from northern New Mexico to the Canadian border. It includes as well the Sierra Nevada and the Cascade Mountains of California, Oregon, and Washington. The topography of this ecoregion is visually stunning but severe and inhospitable. Temperatures can be fiercely cold at higher elevations and higher latitudes; the Northern Rocky Mountains, for example, are much colder than the southern Sierra Nevada Mountains. Precipitation patterns are affected both by topography and by proximity to the Pacific Ocean. Air moving east off the Pacific rises to pass over the successive mountain ranges. It cools as it rises, reducing its capacity to hold water. As a result, the western slopes of these ranges receive considerably more precipitation than the eastern slopes. (This is known as the "rain shadow" effect.) Given this combination of forces, the climate ranges widely between the arid south and the cold and humid northern Cascades. Parts of the Cascade Mountains receive over ten times the annual precipitation of, say, the San Luis Valley in the southern Rocky Mountains.

Much of the ecoregion has been glaciated multiple times, with small but active glaciers surviving in the north. This explains the striking cliffs and hanging valleys of Yosemite and Glacier National Parks, the extensive pattern of alpine lakes, and the thick glacial deposits in the valley floors. Soils tend to be thin and nutrient poor. Dry valleys are dominated by sagebrush and some juniper. At higher altitudes, the vegetation shifts to conifer forests of pine, spruce, and fir, with shrubs, forbs, and grasses covering the forest floor. Above tree line, you'll see alpine meadows of lichen, grasses, sedges, and forbs, with some gnarled bristlecone pine mixed in. Plant diversity is mirrored by animal diversity, with foraging herbivores and a wide range of predators, birds, reptiles, and even amphibians (in the wetter parts of the region).

Through the eyes of European Americans in the nineteenth century, agricultural settlement here was extremely difficult and limited primarily to livestock operations. Settlement occurred primarily around specific natural resources. Mining towns went up overnight around deposits of gold, silver, copper, and other valuable minerals, often dying just as suddenly when the deposits waned. Logging and mill towns grew on the western slopes of the Cascade Mountains in the mid-nineteenth century, reflecting both the richness of the northwest forests and the ease of transportation at ocean ports.

The net result is that this ecoregion saw significant natural resource development but only limited settlement during the nineteenth century. People came to this region but generally did not stay. Private settlement was limited, allowing the federal government to retain much of the region's land. Today the federal government owns 59% of the ecoregion, a much larger proportion than in the ecoregions to the east. Most of this federal land is today managed as national forests and national parks, giving the federal government significant opportunities to shape the ecoregion's health as well as its future.

Ecoregion 10 (North American Deserts)

The North American Deserts ecoregion may lack the conventional beauty of the Northwestern Forested Mountains, but it is no less striking. This region was stretched by tectonic forces, rather than compressed like the mountains to the east and west. As the land stretched, faults emerged, running north and south. Between these faults, vast blocks of rock material tilted dramatically. To roam this region is to see the results. Much of the region, notably central Nevada, is dominated by a relatively uniform pattern of mountains and valleys, all oriented north to south.

Because it sits in the rain shadow of mountains to the west, the North American desert is a mix of true desert and steppe climates. Temperatures regularly exceed 100° F at lower elevations. The soils are dry, thin, and nutrient poor and support very limited biomass. Plants and animals that flourish in these hot, dry conditions have unique adaptations for water conservation and temperature management. Grasses, shrubs, and cacti dominate the landscape. In the north, sagebrush and bunch grasses cover the land; in the south, they are joined by creosote bush and succulents like Joshua trees. The region has a diversity of mammals, though the populations are resource limited in both food and water. Birds of prey are common, and a variety of reptiles flourish in the desert conditions.

For nineteenth-century European Americans, this landscape was foreboding, and even early preservationists like John Muir were rarely drawn to the region. With so little water, timber resources were scant and crops could not survive without extensive irrigation. Agricultural settlement was almost entirely limited to domestic livestock operations that were established near springs and other water sources. The primary draw was mineral resources (copper, gold, iron, nickel, silver), and although prospecting increased as the century progressed, the region was only sparsely populated by the end of the 1800s. Today, the federal government owns and manages 50% of the region as a whole. Federal ownership of this ecoregion reaches its highest level in Nevada, where the government owns roughly 82% of all surface land.

Ecoregion 11 (Mediterranean California)

California contains the only area in North America with a Mediterranean climate, characterized by hot, dry summers and mild winters with variable precipitation. The coast itself receives moisture from sea fog, and the coastal mountains support forests and associated wildlife. The valley floors are hotter and drier, with sagebrush, manzanita, oak, and grasses. The region's vegetation and wildlife thus vary considerably with elevation and proximity to the ocean. The entire ecoregion is essential for migratory birds on the Pacific Flyway, which runs north and south alongside the Pacific coast.

To European Americans in the nineteenth century, the region held significant value, and Native American tribes were quickly displaced. Spanish settlers established large ranches and missions in the eighteenth and nineteenth centuries, long before the federal government acquired the area by treaty with Mexico in 1848. Just as the land was acquired, gold was discovered, setting off one of the most famous mineral rushes in history. More than 300,000 prospectors came to California virtually overnight, giving it sufficient population for statehood by 1850. Thousands of Native Americans died as a result of the European diseases and direct violence that accompanied these arrivals.

The combination of these factors—the dispossession of Native Americans; early Spanish and Mexican settlement; and extensive immigration during the Gold Rush—led to the privatization of much of the area's land in the period before Congress became earnest about permanent land ownership. As a result, the federal government today only owns about 15% of the land in this ecoregion, a proportion which is more similar to the eastern half of the United States than to the other regions of the western half.

Ecoregion 7 (Marine West Coast Forest)

Present-day California also contains the southernmost reaches of a narrow but important ecoregion labeled "Marine West Coast Forest." These forests stretch from the San Francisco Bay to the western tip of Alaska. This is the wettest ecoregion in the country, with parts of California and Washington receiving over 140 inches of precipitation each year. Temperatures are moderated by the Pacific Ocean and rarely drop below freezing at sea level, although at higher elevations, there is enough snow to sustain glaciers as far south as Washington state's Olympic Peninsula.

The topography of this region is mountainous throughout, which in turn correlates with wide-ranging plant and animal communities. Temperate rain forest occupies lower elevations from the Pacific Northwest to Alaska. There, you'll find the tallest tree species in the world (coastal redwood, *Sequoia sempervirens*) and the largest species of spruce (Sitka spruce, *Picea sitchensis*), along with cedars, Douglas fir, true firs, and hemlocks. Epiphytes (plants that grow on other plants) like mosses, ferns, and lichens cover the trees. The sheer biomass of these forests is stunning. At higher elevations, boreal forest gives way to alpine conditions, where a short growing season, heavy snow loads, and extremely thin soils all but eliminate woody plants. Mammals generally associated with American forests (deer, bear, elk, wolves, and smaller mammals like marmots) flourish here; migratory and resident birds abound; and the marine environment contributes anadromous fish.

Native American and Alaskan Native tribes lived throughout this ecoregion prior to European American immigration, but non-Natives settled here long before the United States was formed. Russian fur traders reached Alaska in the 1740s. They hunted and trapped along the coast and established settlements there. Spanish and British explorers arrived in what is now the Pacific Northwest in the late eighteenth century, and European fur traders quickly followed. By the time the Lewis and Clark expedition reached the Pacific Ocean at the mouth of the Columbia River in 1805, there were already well-established European outposts. By the end of the nineteenth century, Native Americans in the southern portion of the region had been largely dispossessed of their land.

The richness of the natural resources might lead you to suspect that private ownership of land dominates this region. But the Russians, Spaniards, and later European Americans that came to these marine forests were largely focused on harvesting natural resources like fur. They did not typically acquire land for agriculture or extensive settlement. Thus the federal government retains ownership of 55% of the region's land.

Ecoregions 2 (Tundra) and 3 (Taiga)

Alaska. The very name conjures up images of remoteness and extremity. Even the sheer size of Alaska is extreme. In the maps in this book, we display Alaska at the same scale as the Lower 48 states so that you can appreciate just how large it is. The Alaskan tundra—the treeless northern area where the soil is permanently frozen—is profoundly foreboding, experiencing months of complete darkness and deadly winter temperatures. Snow remains on the ground for up to 10 months of the year. Trees cannot survive these conditions, so ground cover is dominated by lichens and herbaceous plants. Still, the fauna is remarkably diverse. Bear, caribou, musk ox, moose, wolf, and fox can be found here. The tundra plays a critical role for migratory birds as well, which roost in the region each summer.

There was very little non-native interest in the Alaskan tundra in the nineteenth century, except for Russian fur traders along the west coast and the Aleutian Islands. In the mid-twentieth century, oil companies began developing the Prudhoe Bay oil field, but these companies seldom purchase surface rights to land. Today the federal government owns 63% of the ecoregion—including the 23-million-acre National Petroleum Reserve—and most land not owned by the federal government is owned by the state or Alaskan Natives.

South of the tundra is the taiga, the boreal forest. The word "boreal" simply means "northern" (Boreas was the Greek god of the north wind). In the United States, taiga only occurs in Alaska, straddling the Arctic Circle between the Alaska Range to the south and the Brooks Range to the north. Rolling hills and plains predominate, covered by forests of spruce, poplar, birch, and aspen. Also common are woody shrubs such as willow and alder, and ground covers like lichens, grasses, and sedges. All of these plants are adapted to a very short growing season. Forest mammals include moose, caribou, bear, wolves, lynx, and snow hare. The taiga ecoregion was sparsely populated by Alaskan Native tribes in the nineteenth century, with extremely limited non-native settlements on the southern and western coasts. It remains sparsely populated today. Less than one percent of the surface land in Alaska is in conventional private ownership by non-natives. The remainder is primarily owned by the federal government (55%) or by Alaskan Native Corporations (44%).

Ecological Patterns and Federal Public Lands

Now that we've completed our brief biogeographical tour, look at Figures 1.3a and 1.3b. You once again see the general ecoregions of the Lower 48 states and Alaska. This time, however, the shading represents the percentage of each ecoregion that is federally owned. The federal government owns only a small percentage of the lightest shaded regions, namely the Great Plains and Eastern Temperate Forest. The darkest regions of the map have extensive federal lands. As we've seen, the ecoregions that are darkly shaded include much land that was inhospitable to permanent settlement. We'll return to these ecological patterns later in the book when we discuss some of the key issues of federal land management. Think about federal efforts to protect biodiversity, for example. The federal government has many more opportunities to protect habitat for vulnerable species in ecoregions where it is the dominant landowner than it does in ecoregions where it has only a minor stake. Or consider climate change. Many areas of federal land were set aside as parks and forests because Americans valued their flora and fauna, but what will happen when climate change shifts the boundaries of these ecoregions?

Although we've barely discussed the history of European American settlement in North America, you may already be able to detect the basis for the various classifications of federal land that exist today, from national parks to national forests to wildlife

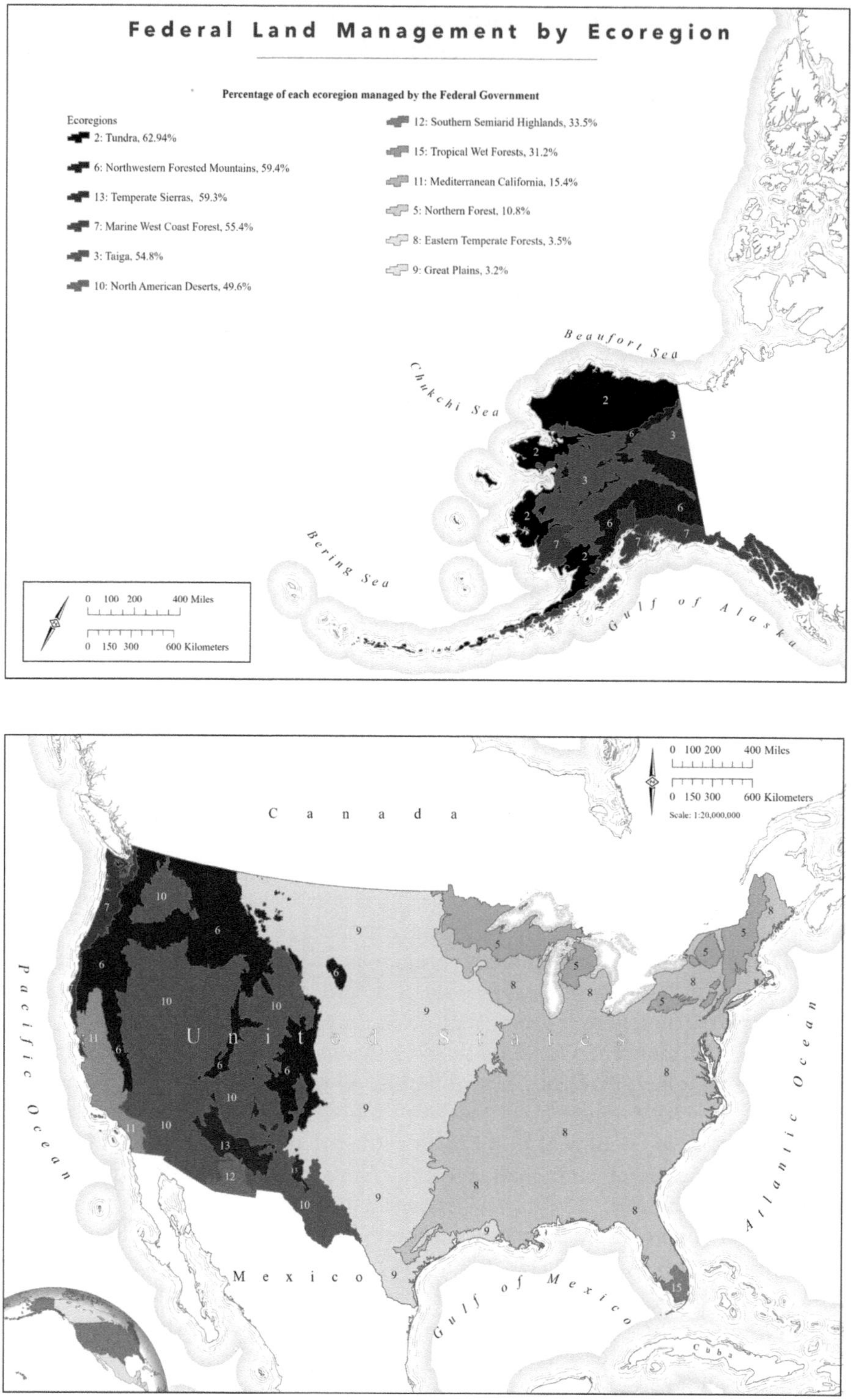

Figures 1.3 a (Alaska) and b (continental U.S.). Federal land management by ecoregion. Ecoregions are shaded by percentage of federal ownership. Maps by Addison Arendt.

refuges. When Congress began to reserve land from settlement in the late nineteenth century, its reasons varied with the nature of the land. First, Congress was interested in retaining areas of exceptional scenery, which were typically areas that had dramatic geological features or remarkable flora and fauna. Many of these areas became national parks. Second, Congress was interested in retaining forested land for perpetual timber harvesting and watershed protection. The lands set aside for these purposes built the national forest system, and it should come as no surprise that the national forests are located disproportionately in the Northwest Forested Mountains and the Marine West Coast Forest. Third, Congress set aside critical habitat for fish and wildlife, especially birds, which formed the National Wildlife Refuge System. Finally, this left considerable acreage that settlers and resource developers did not seek to own and for which the federal government had no clear purpose. Almost all of this land in the Lower 48 states was arid. Today, these lands are managed by the Bureau of Land Management as the National System of Public Lands.

The Plan of the Book

We've written this book to introduce you to U.S. public lands. The book can be divided into three sections. The first (Chapters One through Four) provides an overview of the land itself, including the history of the public lands and the overarching legal structure in which these lands exist. The next section of the book, Chapters Five through Eight, describes the four principal federal land management agencies (the National Park Service, U.S. Forest Service, Bureau of Land Management, and Fish and Wildlife Service) and the systems of land they oversee. The final section (Chapters Nine through Thirteen) works through a number of specific issues that these agencies must confront, from climate change and biodiversity to recreation and wildfire. The following paragraphs provide some additional detail on these chapters.

The first section comprises four chapters. Here in Chapter One, we've explained the biogeography of the public lands, moving east to west across the North American landscape, the same direction of travel in which European settlement proceeded across the continent. Not all land was equally suitable for settlement, and the distinctions revealed in this chapter lay the groundwork (no pun intended) for Chapters Two and Three. Chapter Two describes the process of land acquisition and sale in the early United States, and you'll notice immediately that arable land is the first to pass from government to private ownership. Chapter Three explains how the land that remained in the public domain was classified, and how lawmakers began to see the wisdom in retaining certain land in permanent public ownership. Here is where the idea of the national park was born, along with the national forest, the wildlife refuge, and so forth; lawmakers also created federal agencies to manage these lands. Chapter Four then explains the legal framework that applies to these agencies and their lands. Federal law includes a set of major environmental laws that require land management agencies, among other things, to protect wilderness, to spare endangered species, and to study the environmental impact of their actions.

In Chapters Five through Eight, those agencies are described in detail. Chapter Five covers the National Park Service, an agency within the Department of the Interior that has a unique and important mission. Its purpose is more focused than the other agencies we'll cover, thanks to the legislation that created it—legislation that emphasized the twin goals of preservation and enjoyment of the lands in the National Park system.

Chapter Six introduces the National Wildlife Refuge System and its overseer, the U.S. Fish and Wildlife Service (FWS). In addition to a wide array of small-ish properties along North American flyways, the FWS helps to manage some truly enormous areas in Alaska. Chapter Seven moves on to the U.S. Forest Service, an agency within the Department of Agriculture. The system of National Forests had an unusual and political birth, and its origins and founding mandate continue to shape its management today. Finally, Chapter Eight focuses on the Bureau of Land Management, sometimes called "the nation's largest landlord"—the manager of hundreds of millions of acres across the American West. Although little known and little understood, the BLM plays an exceedingly important role in the management of the mineral and energy resources of the public lands.

Finally, Chapters Nine through Thirteen walk through a set of pressing policy issues related to the public lands. The goal of these chapters is to help you think through these issues in a structured way, just as federal land managers are required by law to do. Chapter Nine begins with the weighty issue of climate change. It explains how changes in climate affect everything from the habitat of endangered species to fire and flood risk, posing enormous challenges for federal agencies. Chapter Ten looks at wildfire directly. Wildfire has emerged as an annual crucible for western lands, and firefighting now dominates the Forest Service's budget. Chapter Eleven takes up biodiversity, exploring the processes dictated by the Endangered Species Act and other federal laws. In Chapter Twelve you'll consider the numerous and fast-evolving forms of recreation that take place on public lands, along with the countless difficulties that can be presented by recreational activities. Chapter Thirteen deals with mining and energy-related activities on the public lands, from hardrock mining to the development of oil, gas, wind, and solar resources.

By the book's end, you will have confronted a wealth of law, history, geography, and public policy, peppered with examples and anecdotes that will, we trust, keep the subject matter accessible and realistic. Along the way, you'll be presented with unanswerable questions and vexing dilemmas but also a toolkit you can use to analyze the various challenges and opportunities facing the U.S. federal public lands. Whether you are considering a career in land management or simply pursuing an interest in this unique American heritage, we hope this volume will help you appreciate the fascinating and important legacy of the public lands.

Resources for Further Study

At the end of each chapter, we'll provide a short list of books and articles that we've found particularly helpful for understanding the chapter's subject matter. The list is not intended to be a complete bibliography, but if you'd like to read further, these recommendations will get you started. Since this chapter is a general introduction, the resources below are comprehensive sources that provide wide coverage of the public lands from a handful of different perspectives.

Eagle, Josh, James Salzman, and BartonThompson, Jr. *Natural Resources Law and Policy* (St. Paul, MN: Foundation Press, 2017).

Gates, Paul Wallace. *History of Public Land Law Development* (Washington, D.C.: Arno Press, 1979).

Leshy, John D. *Our Common Ground: A History of America's Public Lands* (New Haven, CT: Yale University Press, 2021).

Skillen, James R. *This Land Is My Land: Rebellion in the West* (New York, NY: Oxford University Press, 2020).
Sowards, Adam M. *Making America's Public Lands: The Contested History of Conservation on Federal Lands* (Lanham, MD: Rowman & Littlefield, 2022).
Wilkinson, Charles F. *Crossing the Next Meridian: Land, Water, and the Future of the West* (Washington, D.C.: Island Press, 1992).

Notes

1 Emily Holden, "Staff for 'No. 1 public lands enemy' moves to CEQ," *Greenwire*, June 21, 2017.
2 For more about the ecoregions depicted here, see Commission for Environmental Cooperation, *Ecological Regions of North America: Toward a Common Perspective* (1997).

2 How Did the Public Lands Become the Public Lands?

Introduction

This chapter will explain how the public lands came to be. You'll learn how the federal government acquired the nation's land, first by gaining sovereignty from other imperial powers through purchase or conquest, and then by ruthlessly dispossessing Native Americans. The chapter goes on to describe how the government sold off or gave away much of its land after acquiring it. Much of the most desirable land was sold to settlers, granted to newly formed states, or given to railroad companies, but the process was often disorderly and lawless. As the 1800s reached their end, the federal government began to set aside some lands for specific purposes, while also retaining the "leftover" land that was still unclaimed. It is these lands—a mix of lands chosen for particular ends and lands largely deemed undesirable—that form the bulk of today's public lands.

Setting the Stage

Take a moment to look at Figure 2.1, a map of the U.S. public lands. Notice that public ownership seems uneven, disproportionately concentrated in the eleven western states and Alaska. If you zoom in closely, as in Figure 2.2, the map becomes even more chaotic. Public and private lands are interspersed; the boundaries are uneven and irregular. Why are *these* particular lands the public lands?

As explained in the previous chapter, ecological forces inscribed one set of patterns across the American landscape. Indigenous groups created additional patterns through fire, trade, and warfare. But the early American government established a legal framework that created an entirely new set of patterns—patterns that paid little heed to ecological and Indigenous arrangements. Indeed, in the 1800s the federal government actively erased some existing human and ecological patterns. Indigenous peoples were forcibly displaced. The land claims of other European powers were defeated. And all the while, the federal government invested in transportation infrastructure designed to domesticate the formidable topography of the unfamiliar West.

Conventional histories narrate the westward expansion of the young United States in triumphalist and patriotic terms, but the government's acquisition of land was often contentious and violent. And no sooner did officials acquire land than they began to sell or negotiate it away, drawing contorted lines across the landscape in the process. The participants in these affairs included surveyors and outlaws, farmers and adventure-seekers, impoverished settlers and wealthy speculators, hucksters and entrepreneurs. The process of disposal was fast and furious, as was the campaign of acquisition that preceded it.

DOI: 10.4324/9781003392514-2

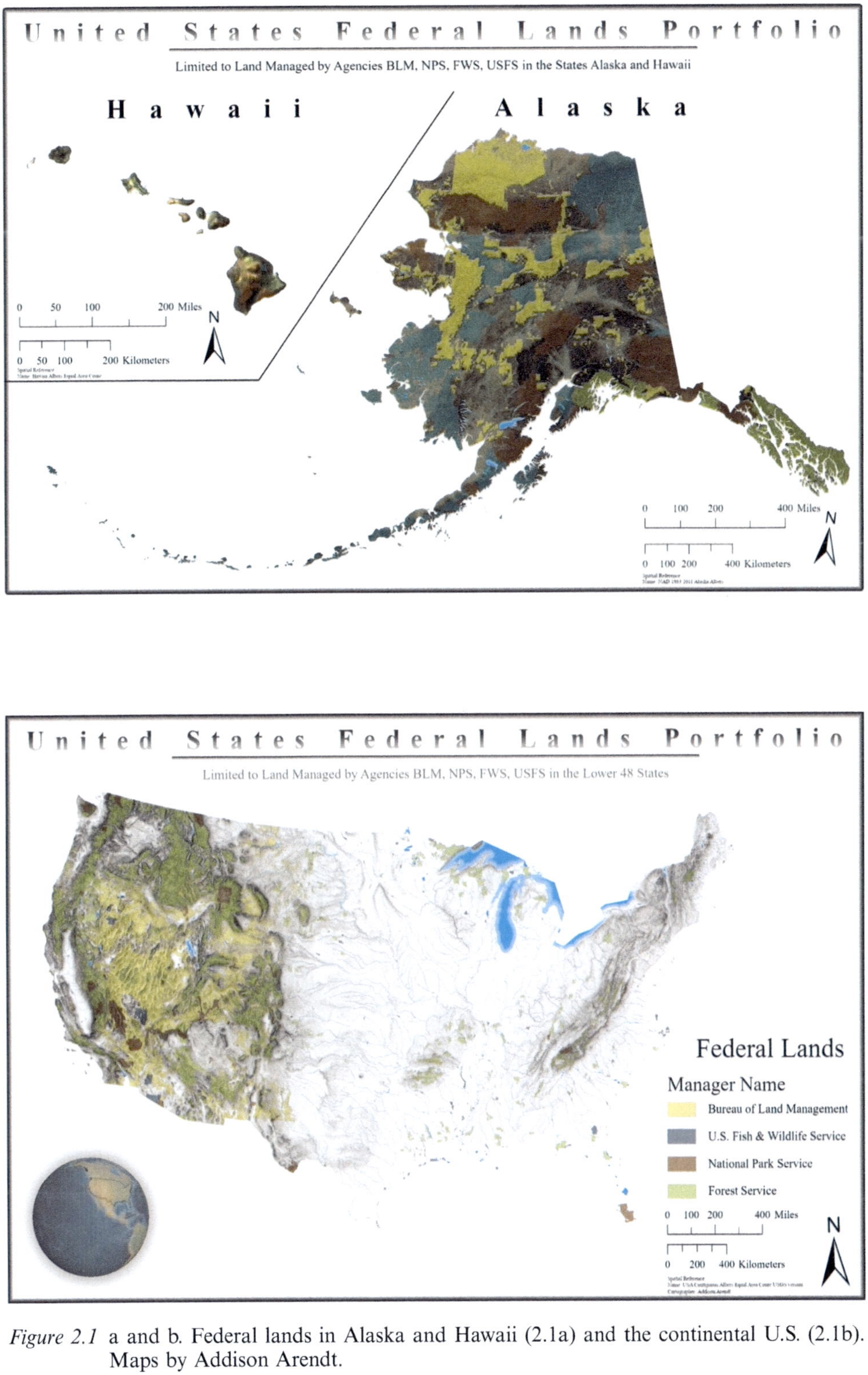

Figure 2.1 a and b. Federal lands in Alaska and Hawaii (2.1a) and the continental U.S. (2.1b). Maps by Addison Arendt.

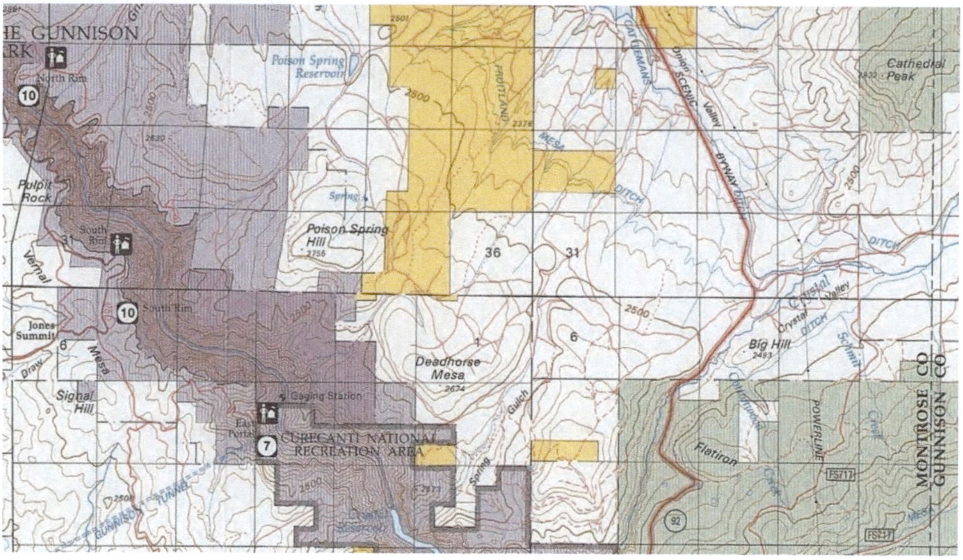

Figure 2.2 Somewhere in western Colorado. The colored areas are various types of federally owned public land. Source: Bureau of Land Management.

The following sections will describe these processes of acquisition and disposal. Sometimes they were careful and deliberate. Often they were hasty and even corrupt. In all cases, they reflected the untidy forces and fragile structures of American democracy. Land ownership is, at one level, a legal matter, the stuff of maps and title documents. But patterns of ownership also resulted in physical patterns on the landscape as owners used their property for agriculture, logging, mining, grazing, recreation, and countless other activities. These patterns, established to a large extent in the nineteenth century, are evident across the federal public lands even today.

Acquisition: How the West Was Won, Bought, and Stolen

How did the United States come to acquire its territory in the first place? Before Europeans colonized North America, the continent was occupied by Indigenous tribes. All of our present-day public lands, in a sense, belonged to them. Pause and reflect on that for a moment: we often think of public lands as areas of pristine natural purity, to be preserved from human influence (as the "Leave No Trace!" mantra suggests). But as Mark David Spence has written, "uninhabited wilderness had to be *created* before it could be *preserved*."[1] The process of creating a pristine "uninhabited wilderness" included many deeply troubling episodes in which Indigenous Americans and unwelcome immigrants were removed, excluded, abused, and killed.

What is perhaps most troubling about this history of dispossession is that much of it was considered legal under both international and domestic law. Many have the impression that colonists simply used raw force to move Native Americans elsewhere, and while there is much truth in that impression, it isn't the full story. European law distinguished *sovereignty* from *ownership*. Under the so-called "doctrine of discovery," explorers were authorized to claim "discovered" territory in the name of their sponsoring state. Such claims were claims of sovereignty—jurisdictional authority—rather than ownership. European legal theorists generally regarded Indigenous tribes as too primitive to claim sovereignty, although most acknowledged that Native peoples remained capable of land ownership.[2] Acquiring the billions of acres that make up the present-day United States, then, required a two-stage process. First, the U.S. federal government had to gain sovereignty over land initially claimed by other imperial powers—France, Britain, Spain, and Russia. Second, the U.S. had to acquire title to its newly claimed lands, which were still occupied by Native peoples.

The First Stage: Acquiring Sovereignty from Other Imperial Powers

Today, we take for granted the stability of most international boundaries, but North American boundaries remained in flux until the mid-1800s. After gaining independence, the U.S. fought, purchased, and negotiated its way across the continent in large, awkward steps (see Figure 2.3). In the massive Louisiana Purchase of 1803, President Thomas Jefferson bought from Napoleon the right of sovereignty over much of the western Mississippi Basin. In the 1840s, the U.S. and Britain signed the Oregon Treaty, formally adding to the United States the land now covered by Washington, Oregon, and Idaho. Around the same time, the U.S. fought a war with Mexico; the war and ensuing negotiations gave the U.S. most of the land from California to Texas. And the

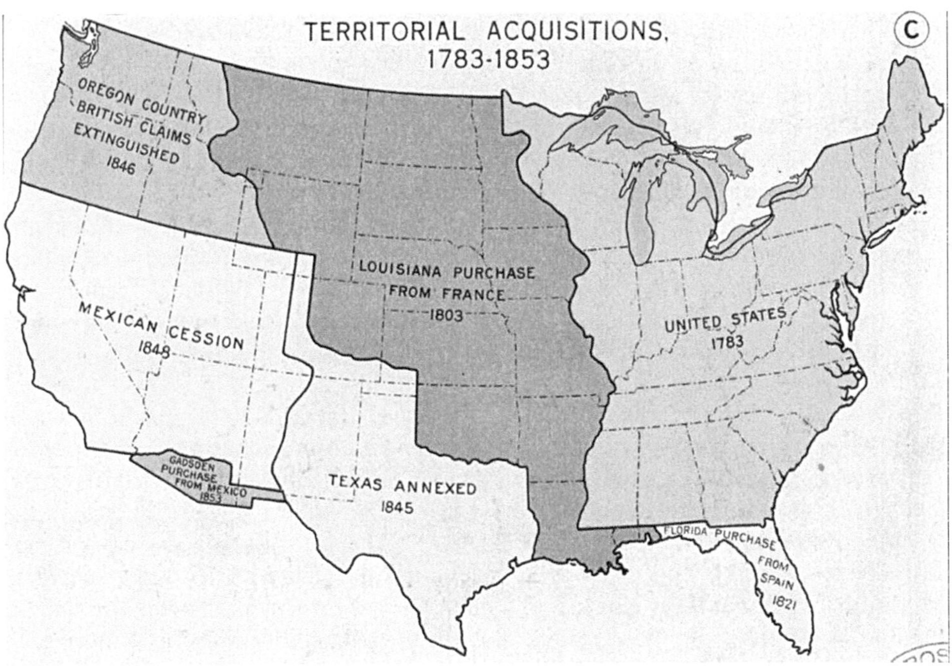

Figure 2.3 U.S. Land Acquisitions During Westward Expansion. Source: Carnegie Institute of Washington, Atlas of the Historical Geography of the United States (1932).

purchase of Alaska from Russia, just after the Civil War, added more than half a million square miles to U.S. territory.

Each of these acquisitions required rigorous negotiation. Most of them involved tense national debates over how and whether to proceed. Like today, Americans held strong opinions—opinions about whether the country was growing too quickly or too slowly, about the budget, about national defense, about Indian policy—and disputing factions often had to compromise before a deal could be struck. But by way of these moves, the U.S. took the shape it has today and, in negotiation with its neighbors, formed its national boundaries.

The Second Stage: Purchasing Title

Again, sovereignty (the right to govern) is not the same as ownership—so, after negotiating the Louisiana Purchase or the Mexican Cession, the U.S. still had to address the question of ownership. Some of the newly acquired land was occupied by French or Spanish settlers, some by American settlers, and much by Native American tribes. These people regarded themselves as owners even though the identity of the sovereign had changed. For example, the Louisiana Purchase included the thriving city of New Orleans. The residents of New Orleans didn't lose their homes due to the Purchase. Landowners in New Orleans were now simply subject to the political authority of the U.S. rather than France.[3]

The same is true of land occupied by Native Americans: if the U.S. wished to own it, the government had to purchase it. Here is a useful passage written in 1947 by Felix Cohen, the leading scholar of Federal Indian Law:

> What we acquired from Napoleon in the Louisiana Purchase was not real estate, for practically all of the ceded territory that was not privately owned by Spanish and French settlers was still owned by the Indians, and the property rights of all the inhabitants were safeguarded by the terms of the treaty of cession. What we did acquire from Napoleon was not the land, which was not his to sell, but simply the power to govern and to tax …
>
> It may help us to appreciate the distinction between a sale of land and the transfer of governmental power if we note that after paying Napoleon 15 million dollars for the cession of political authority over the Louisiana Territory we proceeded to pay the Indian tribes of the ceded territory more than twenty times this sum for such lands in their possession as they were willing to sell.[4]

Twenty times! These purchases, which took place throughout the nineteenth and twentieth centuries, generally undervalued the land and sometimes involved duress or chicanery. Nor could the tribes bargain among multiple buyers: American law, following British precedent, prevented tribes from selling land to anyone other than the federal government.[5] There are remarkable and tragic accounts of American officials and diplomats crisscrossing the West to negotiate land deals with myriad tribes; these accounts, though not the subject of this book, are well worth your time. The map in Figure 2.4 shows how Washington State was purchased from the tribes; each numbered region represents a different treaty.

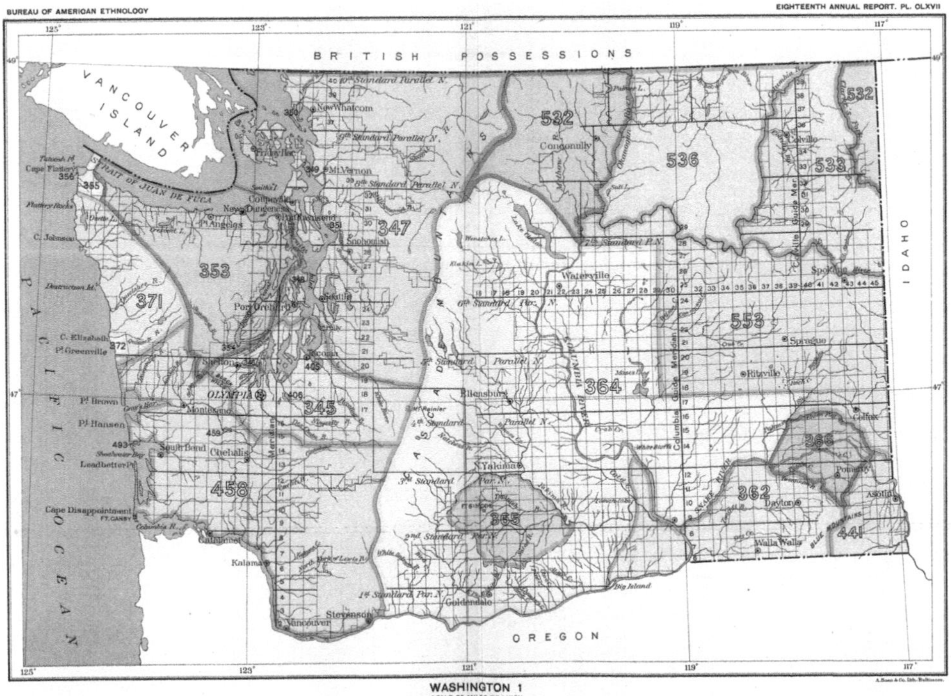

Figure 2.4 An 1899 map of the state of Washington, showing the boundaries of tribal land cessions. Source: Library of Congress.

BOX 2.1 The Lakota Territory

Treaties and land cessions might sound like tidy real estate transactions, but the truth is far more complex and sometimes deeply disturbing. One of the more contentious examples is the federal acquisition of the Lakota territories. The U.S. signed the Fort Laramie Treaty with the Lakota Nation, in which it recognized Lakota ownership of what is now South Dakota and Nebraska along with parts of Kansas, North Dakota, Montana, and Wyoming. But when gold was discovered in the Montana Territory, the federal government violated the treaty. The Lakota, along with the Cheyenne and Arapaho nations, successfully fought to defend their land, leading to a second Fort Laramie Treaty in 1868 that reaffirmed Lakota ownership. But when gold was discovered in the Black Hills in the 1870s, the U.S. once again ignored its treaty obligations and demanded that the Lakota sell their land. When that failed, the U.S. launched a three-pronged assault, which the Lakota and their allies repelled. This included the absolute defeat of General George Custer's cavalry in the valley of the Little Big Horn River. The U.S. responded with force, destroying Lakota and Cheyenne villages and all their inhabitants. The Lakota finally surrendered. Congress then passed a law in 1877 that stripped the Lakota of most of their land in open violation of U.S. treaty obligations.

In the twentieth century, new laws made it possible for the Lakota to sue the federal government for treaty violations, leading to roughly 60 years of administrative review and litigation. In the end, the Indian Claims Commission and several federal courts recognized the federal government's land seizure as an exercise of eminent domain and ordered the government to pay compensation with interest. To date, the Lakota have refused to accept payment, arguing that they are lawfully entitled to the land itself rather than money. The awarded funds have now grown to almost $2 billion—a large sum, but one dwarfed by the billions of dollars in gold that have been mined in the Black Hills since the U.S. claimed the land.[6]

Whether through patently illegal means or through land cessions with a veneer of legality, the federal government gained ownership of the western territories under its jurisdiction. Across most of the American West, if you choose a parcel of property and trace its title backwards from owner to owner, you'll find that at one time the United States government purchased the land by treaty from a Native American tribe before granting it to a settler, a state, or a railroad.

In some instances, these treaties remain important for public land management even today. Tribes often retained certain rights to the land they ceded. For example, the tribes of the Pacific Northwest generally retained the right to fish at "all usual and accustomed places, in common with the citizens of the Territory."[7] State and federal courts have affirmed that many such rights to fish remain intact and have not evaporated simply due to passage of time. They can't be wished away, and those who manage public lands must often take such rights into account as they make managerial decisions.

BOX 2.2 Tribal Fishing Rights in Olympic National Park

When Native American tribes signed treaties ceding much of their land in the 1800s, they bargained for various rights in return: land rights to a tribal reservation, for example, and often the right to fish at their historic fishing spots. Those who manage the public lands are legally required to honor those rights. Below is a short excerpt from the Code of Federal Regulations that explains how Olympic National Park must handle fishing rights. Don't be daunted by the dense legalese; almost every workday, public land managers must consult regulations like this to do their jobs.

36 CFR § 7.28 – Olympic National Park

a. Fishing –

(8) Indian treaty fishing.

i Subject to the limitations set forth below, all waters within the Olympic National Park which have been adjudicated to be usual and accustomed fishing places of an Indian tribe, having treaty-secured off-reservation fishing rights, are open to fishing by members of that tribe in conformance with applicable tribal or State regulations conforming to the orders of the United States District Court.

ii ***Identification cards and tags***. Members of the tribes having treaty-secured fishing rights shall carry identification cards conforming to the requirements prescribed by the United States District Court and issued either by the Bureau of Indian Affairs or the applicable tribe when fishing in accordance with the tribe's reserved treaty fishing right. Such persons shall produce said card for inspection upon request of a National Park Service enforcement officer. A tribally issued identification tag shall be attached to any unattended gear in park waters.

iii ***Conservation closures and catch limits***. The superintendent may close a stream or any portion thereof to Indian treaty fishing or limit the number of fish that may be taken when it is found either that it is:

(A) Reasonable and necessary for the conservation of a run as those terms are used by the United States District Court to determine the permissible limitations on the exercise of Indian treaty rights; or
(B) Necessary to secure the proper allocation of harvest between Indian treaty fisheries and other fisheries as prescribed by the court.

Native tribes usually attempted to reserve some land for themselves in treaty negotiations, hence the term "reservations" for lands currently in tribal possession (more about this later in the chapter). The story of tribal reservations is complex and tragic. U.S. policy toward Indian tribes has swung drastically over the years and includes eras in which lawmakers sought to assimilate Indigenous peoples into the dominant culture or even to "terminate" tribes, which in legal terms meant that the federal government stopped recognizing those tribes as sovereign. To eliminate tribal land ownership, the government would allot tribal land among individual tribal members and sell or give away any "surplus" land. Once the land was unrestricted, outside purchasers quickly reduced tribal presence in the former reservation. Together, termination and allotment struck simultaneous blows against tribal identity. Policy later swung back and attempted to restore tribal reservations, but because so much land had been sold to non-Indians, reservations today contain countless islands of private land that are outside tribal jurisdiction for many purposes. Even beyond the resulting jurisdictional chaos, it's no overstatement to say that under contemporary international law U.S. policy would be condemned for violating human rights.

Disposal

Although this book separates acquisition and disposal for the sake of clarity, the two processes were happening alongside each other. The United States didn't buy land from tribes in order to keep it permanently but to sell it. American settlers and traders were forging their way across the continent, hungry for land—so hungry, in fact, that they often built permanent settlements even before they knew whether the law allowed them. Sometimes the government sold future rights to land before it had acquired that land from native tribes. Settlers, traders, mineral hunters—all were uniformly eager for the government to resolve Indian land claims and to sell the land as fast as possible.

Congress passed thousands of individual statutes authorizing the transfer of land from the federal government to states and private owners. Some authorized regular cash sales, some rewarded military veterans with land in return for their service, some authorized land grants for education, and some authorized land grants or sales for specific purposes from agricultural homesteading to timber and mineral development. The list of examples in Table 2.1 reveals the breadth of Congressional action.

The rest of this section describes the many ways that federal land became private land. It's through these processes that the jumbled lines on our maps will start to make more sense. In an effort to help make an enormous, chaotic history more digestible,

Table 2.1 Federal Land Disposal Laws

Land Sale Statutes	
General Land Ordinance of 1785	Established the rectangular survey, reserved Section 16 of each township for the states to fund education, and authorized sale of the remaining land at auction for no less than $1 per acre.
Land Law of 1796	Authorized sale of land north of the Ohio River for a minimum of $2 per acre in 160-acre or 640-acre parcels.
Land Law of 1800	Authorized sale of 320-acre parcels and gave purchasers 4 years to pay.
Land Law of 1820	Required full payment at the time of sale but reduced the cost to as low as $1.25 per acre and offered smaller, 80-acre parcels.
Graduation Law of 1854	Created a clearance sale for lands that languished on the market. Land that had been open to sale for more than 10 years cost $1 per acre; for more than 15 years, $0.75 per acre; and so on.
Coal Lands Law of 1873	Authorized sale of lands with coal deposits (160 acres for an individual, 640 acres for associations). The price was $20 per acre if it was within 12 miles of a railroad and $10 per acre if it was further from a railroad.
Desert Land Law of 1877	Authorized sale of desert land sections (640 acres) that lacked timber or minerals for $1.25 per acre.
Timber and Stone Law of 1878	Authorized sale of 160-acre parcels valuable chiefly for timber and stone for $2.50 per acre.
Land Grants and Laws Authorizing Land Patents	
General Land Ordinance of 1785	Set aside one seventh of the land for "military land warrants" given to veterans for their service. The amount of land awarded in these warrants varied by military rank.
Internal improvement grants in 1841	Granted 500,000 acres of public land to public land states to support internal improvements.
Railroad grants 1850–1872	Granted 20–40 odd-numbered sections of land for each mile or track built. Between 1862 and 1871, Congress granted almost 130 million acres of land to railroad companies.
Homestead Act of 1862	Authorized individuals to apply for 160 acres and gain title after 5 years of residence or cultivation. (Of roughly 1.3 million homestead applications before 1900, only about half were patented.[8])
Morrill Act of 1862	Granted each state 30,000 acres of public land per Member of Congress to fund agricultural and mechanical colleges.
Lode Mining Law of 1866	Authorized miners to stake claims and patent them if they invested a minimum of $1,000 in labor and/or improvements.
Mining Law of 1870 and Mining Law of 1872	Authorized placer claims of no more than 10 acres and lode claims that could be 1500 feet by 600 feet. Land could be patented after minimal time and investment for $2.50 to $5.00 per acre.
Timber Culture Act of 1873	Authorized grant of 160 acres to any individual who planted 40 acres of trees, no farther than 12 feet apart, on denuded land. Later amendments reduced the requirement to 10 acres of trees.

we've labeled the following subsections according to six of the most important concepts, using three Ss and three Rs: Settlement, Survey, and States; and Resources, Railroads, and Reservations.

Settlement

Simply put, settlement is the phenomenon most responsible for the current boundaries of public lands. Land that was desirable for settlement in the 1800s is largely now in private hands; land that was not settled is, today, quite likely in public ownership. Obviously, that's an extraordinary generalization, and although it captures some important historical realities, it needs qualification. So let's dig a bit deeper.

In the 1800s, Americans proved to be a restless people, pressing west in an inexorable tide. There was great moral and practical energy behind the process of westward settlement. Some people were motivated by poverty or opportunity, some by the prospect of the West's vast untapped potential, some by religious persecution. The federal government needed European Americans to settle western lands in order to demonstrate and secure American sovereignty. In other words, settlement was driven both by the pull of settlers and the push of the national government.

Lawmakers in Congress tried very hard to organize and contain the process of settlement. But the federal government was much smaller and weaker than it is today, so many people traveled west and occupied the land they found long before the law created a system for land acquisition. This was particularly true on or near the settlement frontier, which the U.S. Census Bureau defined as an area with fewer than two people per square mile. Although those who moved ahead of the law were clearly trespassers or squatters, Congress occasionally enacted "preemption" laws that gave settlers a priority right to buy the land they occupied—which would otherwise have awaited sale at auction to the highest bidder.[9] In short, even those settlers who violated the clear terms of the law often received, ultimately, clear title to the land they sought.

Eventually, Congress resigned itself to the reality that settlers were a force not easily contained. Laws like the famous Homestead Act of 1862 changed the model of federal land disposition to conform to the state of affairs on the frontier. Under the Act, settlers could occupy up to 160 acres of unsettled public land and acquire title to it after cultivating the land for 5 years. (For reference, a 160-acre tract of land is equal to a square with sides of one-half mile in length; a single acre is roughly the area of a football field.) Over 250 million acres eventually passed into private ownership by way of the Homestead Act.

What kind of land was attractive to settlers? Mostly, productive farmland. That's why land in the enormous Mississippi River drainage, from Louisiana to Minnesota and Tennessee to Nebraska, was in high demand. Nearly all of this land is now in private ownership. In fact, it was almost entirely claimed by 1900. (Recall from Chapter One that precious few verdant lowlands or grasslands remain in public ownership.)

But as settlers approached and entered the Rocky Mountains and the Great Basin, the dynamics changed dramatically. Rather than trying to direct a flood of settlers, the federal government had to *promote* settlement. That is because as you move from east to west across the 100th meridian—a line of longitude that runs through North and South Dakota, Nebraska, Kansas, Oklahoma, and Texas—annual precipitation becomes too limited for unirrigated agriculture. Settlers moving past this line found that sources of water were few and far between and that natural vegetation was sparse. Sites appropriate for cultivation were difficult to find, and because the land was less productive, standard 160-acre tracts were insufficient to maintain a farming unit. Although Congress modified the rules for land claims to allow 640-acre tracts (one square mile), the physical characteristics of the land still made settlement a burdensome enterprise. By far the most desirable land feature in the arid West was access to water—a spring or a stream, even a seasonal one. Private claims dotted the landscape wherever water was to be found, surrounded by vast dry lands of little interest to settlers. This suited a ranching economy, since ranchers could claim a 640-acre base ranch with water access and then graze livestock freely on the surrounding public domain. Indeed, federal courts were explicit that in the absence of federal prohibitions, the public domain was a commons open for ranchers to use at will. In 1890, for example, the Supreme Court ruled that

> there is an implied license, growing out of the custom of nearly a hundred years, that the public lands of the United States, especially those in which the native grasses are adapted to the growth and fattening of domestic animals, shall be free to the people who seek to use them where they are left open and unenclosed, and no act of government forbids this use.[10]

Even with this bare-bones history, the disfigured outlines of the Western public lands become much more intelligible. We can now begin to see why low-lying farmland is not included in the public lands; why islands of private ownership, containing springs or water sources, lie within many bodies of public land; and why mountainous, arid land, or land that was otherwise difficult to cultivate or irrigate, often remained in public ownership. Moreover, we can now begin to understand why so many western tracts come in increments of 160 or 640 acres. The lines by which those tracts were drawn are our next subject.

Survey

We noted a moment ago that impatient settlers moved faster than Congress, which was desperately trying to channel the torrent of settlement. What tools did Congress have? One of its most important was the land survey. If you've ever wondered why property boundaries in most of the country are straight rather than curved, you've got the survey to thank for that.

Surveying land was a way of systematically exploring it in great detail while also demarcating it in a way that would prepare it for sale. The project of surveying the land west of the original thirteen colonies began in earnest in the 1780s and continued for over a century. In the General Land Ordinance of 1785, Congress instructed surveyors to adopt a new method of surveying called the "rectangular survey" (see Figure 2.6). Survey teams were directed to measure and mark land in one-mile squares called "sections," which were themselves grouped into square "townships," six miles to the side, and numbered from 1 to 36. (Depending on where you live, your township lines may well exist to this day, and may even serve as boundaries for your local government units, including school or library districts.) This is why Congress offered ranchers exactly 640 acres (one square mile) and homesteaders exactly 160 acres (a "quarter section," one-half mile square), and why so many land parcels today are still 640 acres, 160 acres, 40 acres (a quarter of a quarter section), and so on.

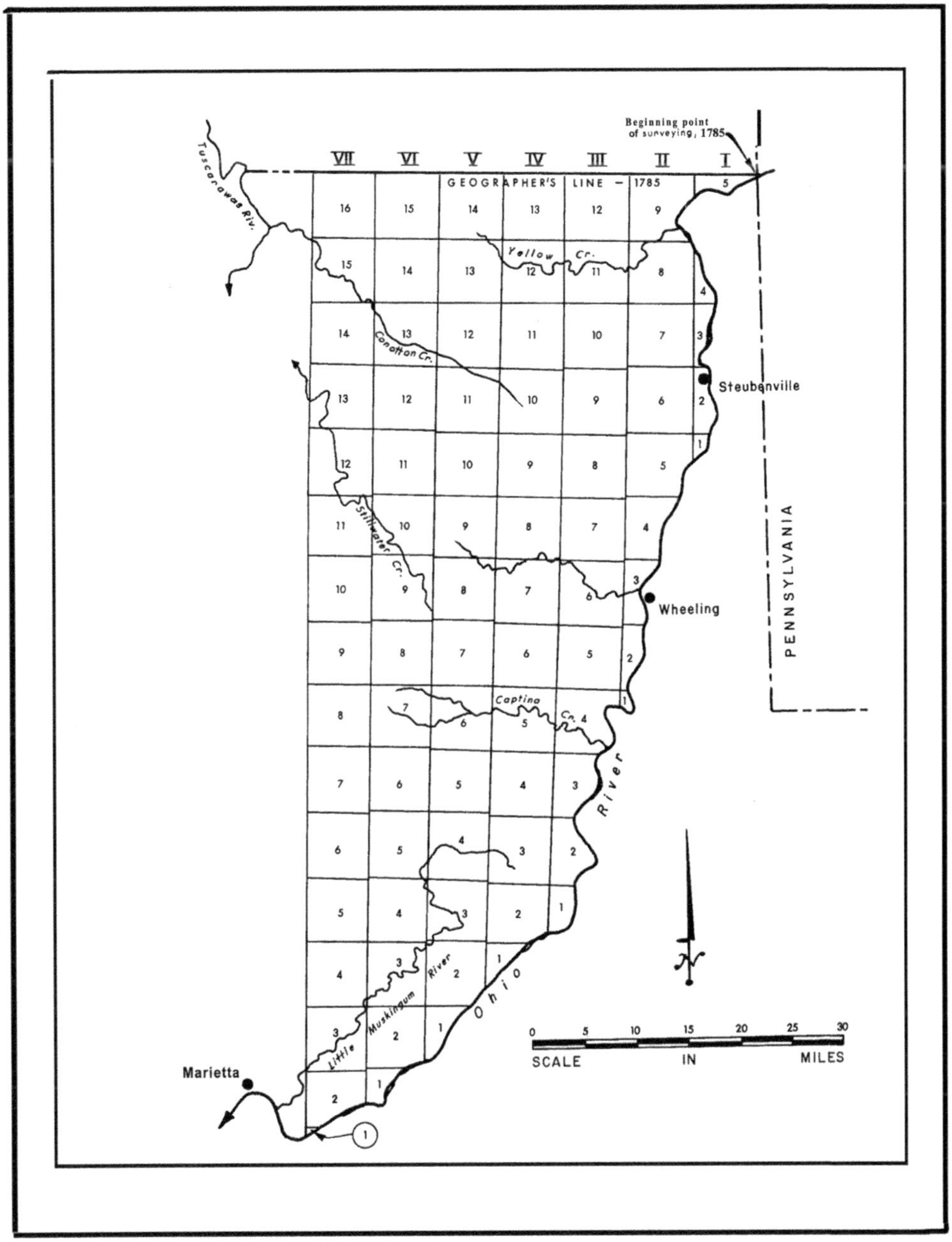

Figure 2.5 A map of the land survey's initial progress. The 1785 starting point was at the top right, where the Ohio River crosses into Pennsylvania. Source: Bureau of Land Management.

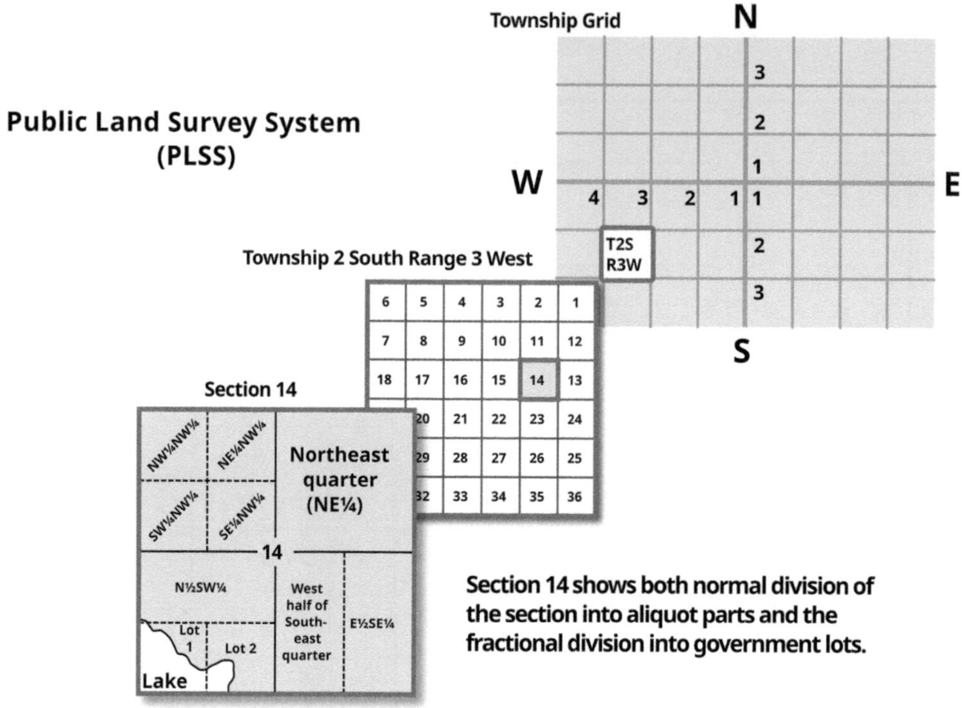

Figure 2.6 The Public Land Survey System. Source: Wikimedia Commons.

We have grown so accustomed to straight, perpendicular property boundaries that it may be difficult to imagine land being demarcated any other way. But before the U.S. land survey, few jurisdictions around the world divided their property by way of a giant grid. Even in the original American colonies, land was typically demarcated by "metes and bounds," an ancient system in which deeds to property identified boundaries with reference to visible landmarks—a massive oak tree, a sturdy fence post, an immovable boulder, a recognizable river bend. To this day, parcels of land in New England often have distinctive, asymmetric shapes, many of which date back to the colonial period.

A straight-line grid system had some distinct practical advantages. It reduced boundary conflicts between neighbors, which had plagued the metes and bounds system ("Hey! Someone moved the big rock and cut down the old tree!"). It created an easy way of identifying parcels in a way that supplied immediate geographic information as well. And perhaps most importantly, by creating equally-sized sections, it commodified land and facilitated standardized pricing and other land programs. Speculators and settlers alike could bid on land without seeing it first, knowing roughly what they would get.

The rectangular survey's advantages were particularly evident as settlers moved west past the Ohio River and across the prairies and plains of what we now call the Midwest. After all, one 160-acre parcel of rich Iowa farmland has roughly the same value as the next, and clear property delineation avoided overlapping land claims.

But the rectangular grid system came with some side effects. Surveyors' grid lines *by design* don't correspond to the physical characteristics of the land. As has been said, "nature abhors a straight line."[11] From the standpoint of ecology, water management, or even agriculture, it might have made more sense for land boundaries to correspond to ridge lines, waterways, or other natural features. The relentless rectangularity of the survey map is not well suited to mountain topography, veins of valuable minerals, or riparian tree stands. In the West, geographer John Wright notes, we find the "[survey] grid overlaid onto topography more suited to calculus than geometry."[12] In short, the rectangular grid was perhaps better suited to Iowa and Illinois than to Colorado and California.

States

So far, we've explained how waves of migrants swept across the country, often out of contact with the distant government of the United States. We've described how the sparse Indigenous population was cajoled and compelled to sell land. And we've now seen why land boundaries in the West and Midwest are usually straight, perpendicular gridlines.

The next topic to explore is the important process by which the territory of the United States was converted into sovereign states, equal in authority and stature to the original thirteen colonies. The balance of power between the states and the federal government remains one of the most important and distinctive attributes of American governance, and it has enormous implications for public lands issues.

When the United States assumed jurisdiction over land from France, Mexico, or Britain, it was assumed that new states would be carved out of that land. The famous Northwest Ordinance of 1787 laid out a process for the formation of new states. The typical pattern was that land was first organized into territories, with governments that were creatures of and subservient to the federal government. (In contrast, state governments are sovereign entities which retain, per the Tenth Amendment to the U.S. Constitution, all government powers not specifically granted to the federal government.) The process by which territories were converted into states was complicated and thoroughly political. Numerous hot-button issues were negotiated. For example, one of the most important questions prior to the Civil War was whether newly admitted states would permit slavery. You may recall that the infamous 1820 Missouri Compromise provided a fragile answer by admitting Missouri as a slave state and Maine as a free state before the Compromise dissolved in the 1850s.

What is less known, however, is that proponents of new states also fiercely negotiated for ownership of federal land within the boundaries of the states-to-be. It was generally accepted that the U.S. would grant at least some land to fledgling states as a source of revenue for schools, roads, and other public infrastructure, especially because the Northwest Ordinance prevented states from taxing land owned by the federal government. When Congress created Ohio in 1803, the new state was granted one section of land per township—Section 16, near the center of the township. The state was free to use that land or to sell it.

BOX 2.3 The Indiana Enabling Act of 1816

The language below is the actual text of the law passed by Congress to create the State of Indiana. Notice that, in this excerpt, Congress granted land to the new state to support education generally and to support a seminary in particular, as well as to provide for the "seat of government," i.e., the state capital.

An act to enable the people of the Indiana Territory to form a constitution and state government, and for the admission of such state into the Union on an equal footing with the original states.

Be it enacted by the senate and house of representatives of the United States of America, in congress assembled, That the inhabitants of the territory of Indiana be, and they are hereby authorized to form for themselves a constitution and state government, and to assume such name as they shall deem proper; and the said state, when formed, shall be admitted into the union upon the same footing with the original states, in all respects whatever ...

Sec. 6. And be it further enacted ...

First. That the section numbered sixteen, in every township, and when such section has been sold, granted or disposed of, other lands, equivalent thereto, and most contiguous to the same, shall be granted to the inhabitants of such township for the use of schools ...

Fourth. That one entire township, which shall be designated by the president of the United States, in addition to the one heretofore reserved for that purpose, shall be reserved for the use of a seminary of learning, and vested in the legislature of the said state, to be appropriated solely to the use of such seminary by the said legislature ...

Fifth. That four sections of land be, and the same are hereby granted to the said state, for the purpose of fixing their seat of government thereon ...

Even after states were formed, they begged Congress for additional grants of land—and, thanks to their new representation in Congress, they often succeeded. Congress granted land to states for roads, for canals, for railways, for universities, and so on. As the United States expanded, advocates for new states, having seen older states win more land, successfully bargained for expanded initial land grants from Congress. A number of states, beginning with California in 1850, were granted not just Section 16 but also Section 36 of each township, creating a pattern like the one visible in Figure 2.7. Utah, which became a state in 1896, was granted *four* sections per township.[13]

Ultimately, states received "princely portions" of the federal public domain.[14] This is unsurprising insofar as every member of Congress is also a representative of a particular state, and the admission of western states altered the balance of power (especially in the Senate, where each state had two senators regardless of population) to increasingly favor the states west of the Mississippi.

What happened to all the land given to states? Some of it remains in state ownership today and is roughly akin to the federal public land. (In fact, sometimes squares of state land punctuate the National Forests or even the National Parks.) But much of it was eventually sold into private ownership. Today, you'll often find tracts of private land that occupy Sections 16 or 36, or that can otherwise be traced back to a grant from the federal government.

Resources

The story of how the public lands took shape can't be told without discussing the incredible natural resources on, under, and around those lands. Precious minerals like gold and copper; necessities like water, timber, or forage; and fuels like wood, coal, oil,

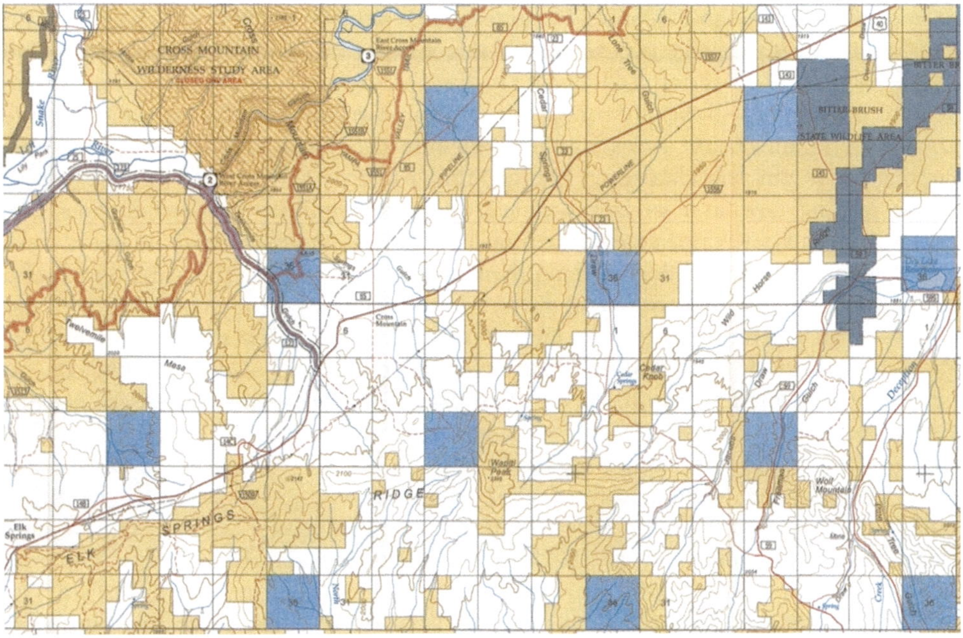

Figure 2.7 Somewhere in Colorado. The blue sections (township sections 16 and 36) are state lands, granted to Colorado at statehood. Source: Bureau of Land Management.

and gas are now and have long been abundant across North America, but especially in the mountainous regions of the American West, where public lands predominate. The endless hunt for such resources drew multitudes to these regions and—like settlement, the survey, and the states—helped give the public lands their shape.

A few pages ago, we made the point that settlers took the land they wanted and that the public lands are essentially the "leftovers." We noted that agricultural land was particularly desirable. So also was land containing fuels and minerals. Today, a good deal of western land rich in those resources is privately owned. But we can draw a useful distinction here between renewable and nonrenewable resources. Land rich in renewable resources, like water or forage or timber, often remained valuable in perpetuity, which encouraged long-term private ownership. Deposits of nonrenewable resources like silver or copper, by contrast, often had no residual value once the deposit was depleted. So today, while many forested areas or lands with good water access remain in private ownership, the public lands are dotted with tens of thousands of abandoned mines and, of course, even the ghost towns of American folklore. Some of this land simply returned to federal ownership by default.

The California Gold Rush is a prominent example of the sort of resource race that took place up and down the mountain ranges of the North American West. The discovery of gold in the foothills of the Sierra Nevada Mountains in 1849 attracted countless prospectors and, behind them, honest purveyors of support services as well as cunning opportunists. All were eager to make a quick buck; all were, in form if not in feeling, trespassers on land owned by someone else. To be sure, it often wasn't clear whose land it was—but it was quite clear, at least to many officials of the U.S. government, that the removal of gold, silver, or even timber from federal land constituted a federal offense. But those officials, far across the continent, couldn't do much about it and found themselves just as powerless in this instance as they were more generally in the larger process of western settlement. Just as Congress eventually allowed settlers to acquire arable land simply by occupying and improving it, so did lawmakers finally grant mineral prospectors a vested right to the mineral deposits they worked. The most famous law to this effect was the General Mining Law of 1872.

The Mining Law, as it is simply called, essentially ratified the informal rules that prospectors had developed among themselves. The Law told prospectors how to "stake a claim" to minerals and then granted them an exclusive right to those minerals, good even against the federal government itself. A successful prospector could go further and obtain a patent (complete title) to the land after meeting a few simple requirements. Prospectors paid only $2.50 to $5.00 per acre for the land and paid no royalties on the minerals they withdrew. In this way a good deal of mineral-rich land passed into private ownership—along with plenty of acres claimed fraudulently by scammers who simply lied about the presence of minerals.

Importantly, the Mining Law of 1872 added a new set of patterns across the public domain that had no relationship to the rectangular survey. Old mining claims remain visible today in the form of oddly-shaped private tracts, scattered amidst adjacent public lands (see Figure 2.8). After all, mineral deposits generally aren't found in tidy square deposits of 640, 160, or 40 acres. Lode mining claims follow veins of mineral ore, veins which pay no heed to the lines of the land survey.[15] Sometimes, lawmakers buy back these inholdings but many remain, testifying to the land rushes of yesteryear.

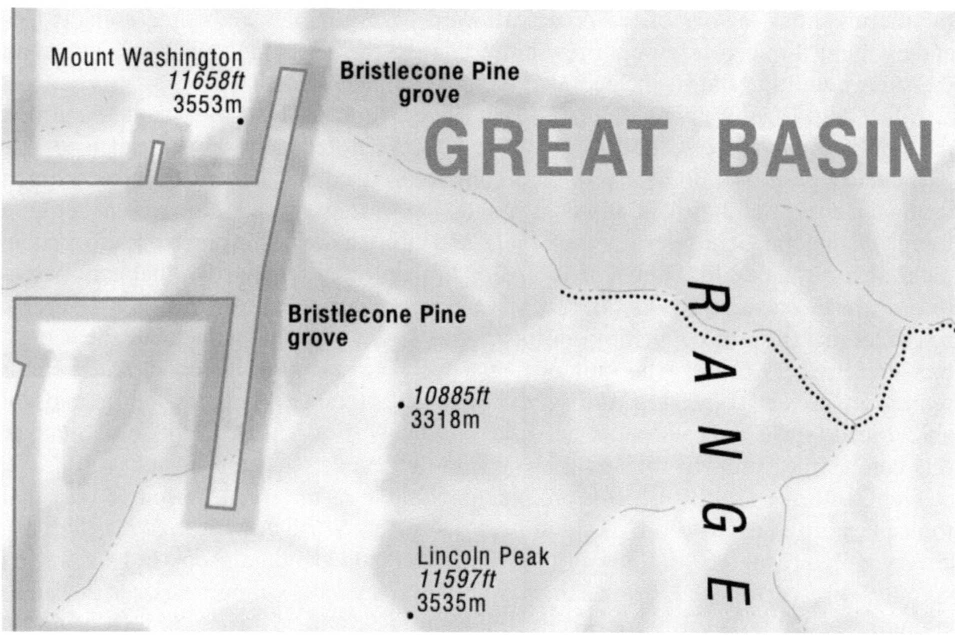

Figure 2.8 A portion of Great Basin National Park in Nevada. Note the unusual shape of the Park boundary's western edge, which reflects pre-existing mining claims not included in the Park. Source: National Park Service.

What is more, the Mining Law did not require prospectors to tell the federal government when they staked a claim; they only needed to approach the federal government if they wished to patent their claim after finding valuable minerals. The result is, in certain parts of the West, a landscape covered in a jumble of mining claims, many of which were never patented but may nonetheless remain valid.

Much, much more could be said about how the physical resources of the land shaped the human presence on that land, both past and present. For now, suffice it to say that the boundaries of today's public lands often gesture toward forgotten stories involving a chase for valuable resources.

Railroads

It may surprise you that railroads earn their own section in this discussion, but the transcontinental railroads were an epic force in nineteenth-century America; their completion was a technological shift as momentous as the telephone or the motorcar. Our concern here is with their mark upon the land, which extended far, far beyond the narrow thread of the railroad tracks themselves.

Railroads were fantastically expensive. They required extraordinary quantities of raw materials as well as human labor, the bulk of which was provided by Chinese immigrants whose exploitation is another hideous aspect of a supposed American triumph. As with many capital-intensive projects, the railroad would require titanic investment before it returned even a modest revenue. The private banks of the time were not up to the task of financing such a massive project, so entrepreneurs turned to the federal government for help. And what help they received!

The federal government had an unsteady cashflow, but what it had in abundance was land. Pioneering railroad firms negotiated with Congress for so much acreage that it almost beggars belief. The idea was that the railroads, like the states before them, could sell their land for the necessary revenue. In short, Congress invited the railroads to chart a suitable path across the country and then granted them not only the strip of land required to construct the tracks (typically a right-of-way about 200 feet wide) but also half of the sections in a broad swath extending 10, 20, or even 40 miles on *both sides* of the rail line (as Figure 2.9 vividly depicts).

Not *all* the land but half of it. Why half? Congress knew that wherever the railroad went, land values would rise, and it sought to retain some of that appreciation for the benefit of the general public. The railroads, in other words, would not be allowed a monopolistic stranglehold on the development that would no doubt accompany the lines' construction. But the particular fashion in which Congress accomplished this grant created a most unusual land pattern that remains glaringly visible across much of today's public landscape: the checkerboard.

The gargantuan land grants to railroads were of *alternating* sections, a seemingly benign decision but one with bizarre repercussions. Like a checkerboard, no two adjacent sections would belong initially to the same party, and an owner of multiple sections would be forced to trespass (albeit minimally) to cross diagonally from one section to another. The resulting system virtually guaranteed that future generations would confront nearly insuperable challenges of land access and management. Congress probably assumed that the U.S.'s retained land would pass quickly into private ownership, but even now, public and private owners are checkered uneasily across vast areas in the West—and conflicts over trespass are certain.

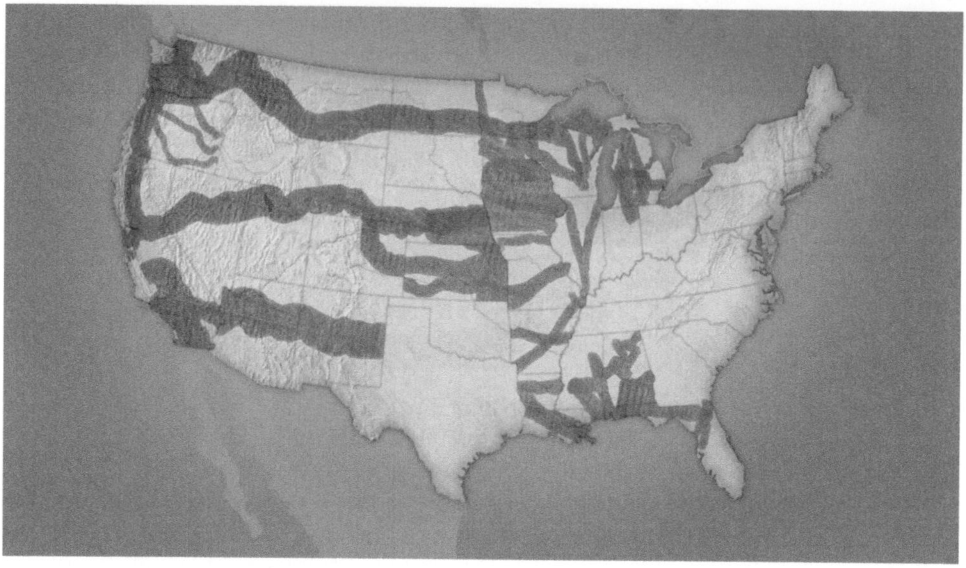

Figure 2.9 Land Grants to Railroads. Railroad companies received alternating square-mile sections inside the shaded areas. Source: Forest Service.

Today, public land managers often find themselves managing lands differently than their neighbors on every side—perhaps with respect to timber cutting, perhaps oil and gas drilling, perhaps cattle grazing (see Figure 2.10). To make matters worse, the U.S. Supreme Court has suggested that members of the public may not have the right to cross private land at the checkerboarded corners, effectively landlocking mile-square sections of public land—land we all share but sometimes can't lawfully access.[16] (The law is more generous to private landowners, who typically *are* allowed to cross public lands to get to their landlocked parcels.) The situation may sound almost comical, but the implications are serious when, for example, a timber company clear-cuts a large area of private forest immediately adjacent to a section of National Forest land—not only impairing the integrity of the forest as an ecological habitat but causing erosion, increasing runoff, and so forth.

In sum, it would be hard to overstate the importance of railroads in shaping new patterns on the landscape. Rail routes spanned the continent several times over by 1890. Checkerboarded swaths, up to 80 miles across, memorialized the nation's colossal land grants to railroad tycoons. The social and economic implications were also profound, as railroads integrated people, land, and resources into the national market economy. Reliable, year-round links now connected the West with cities like New York and Chicago. As historian William Cronon explains in his seminal work *Nature's Metropolis*, railroads reduced the friction of distance.[17] Farmers in Iowa could ship grain via Chicago to markets in the East; logging companies and mills sent lumber to settlers on the prairie. While some goods were still shipped by river, those lines faded in economic significance compared to the railroads.

Reservations

The final concept to discuss here is the *reservation*. We've already mentioned that the semi-sovereign lands of Native American tribes are called reservations, but the term has a much broader application in public lands law. A reservation is any instance in

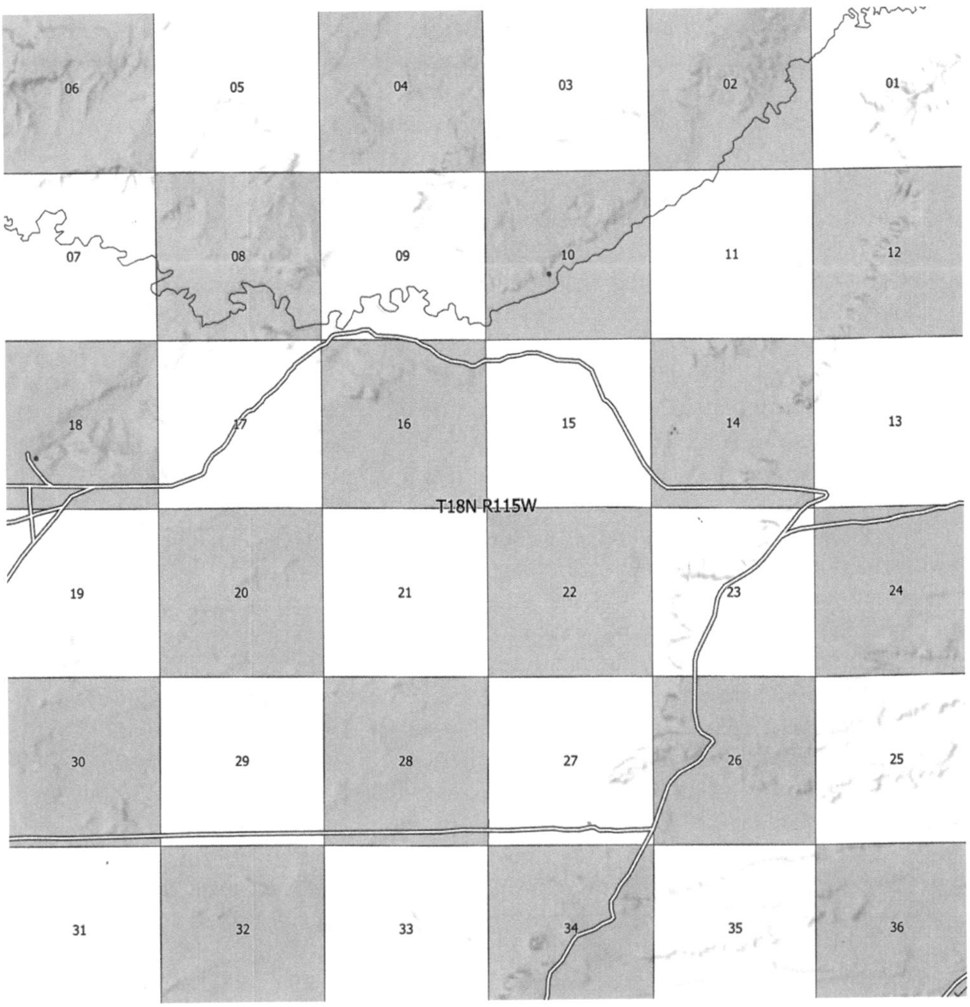

Figure 2.10 Checkerboarded lands in southwest Wyoming. The darker sections are BLM land; the lighter sections are privately owned. Each section is one square mile. Source: U.S. Forest Service.

which government sets aside land for a given purpose, thereby preventing the land from entering private ownership. You'll recall from the beginning of the chapter that, at the outset, Congress didn't intend to reserve much land at all. The assumption seemed to be that any land in public, federal ownership would be quickly sold, conveyed to a state, or otherwise disposed. But by about 1900, that view was changing rapidly. Many Americans supported the idea of reserving land for all kinds of public purposes.

Before we discuss those purposes, let's return to tribal reservations, which remain of fundamental importance across the American West. Remember that prior to European contact, Native American tribes spanned the continent. Under European legal theory, these tribes *owned* what is now the United States—all 2.4 billion acres of it. Through treaty and brutal conquest, the federal government took title to almost all of this land. Tribal reservations are now limited to just 56 million acres, or about 2% of U.S. territory.

Today, these reservations bear the weight of a profoundly unjust history. Native American communities suffer disproportionately not only from economic distress but from disease, inadequate education, and a lack of social services. The way of life enjoyed by their tribal forebears just a few generations ago is utterly unrecoverable, wiped off the land by the swelling tides of settlers, prospectors, and others who saw in the West not pre-existing people but land they could acquire and reshape to suit their own needs. A crucial task for the public servants who now manage our lands is to manage them for all Americans, including (and perhaps especially) for those Native Americans whose historical and spiritual connections to their homelands have been so grievously damaged.

In a sense, the loss of land that is the tragic history of Indigenous tribes led also, and more constructively, to different forms of public land reservation. As the rapacious forces of settlement and occupation spread across the wide continent, a variety of forces began to stir in the American consciousness. For one, the breathtaking beauty of western landscapes, so utterly different from the familiar sights nearer the Atlantic, slowly circulated through the day's media and entered the public imagination. For another, rumors spread of the impending and unimaginable depletion of resource stocks that had seemed inexhaustible only several decades earlier.

These forces were slow to coalesce, but they eventually gave rise to a nascent conservation movement. More about that in the next chapter. For now, the word "movement" is far too grandiose for the few small steps undertaken in the 1800s that are here in view. Such steps include the decision by Congress to set aside a portion of northwestern Wyoming, called Yellowstone, as a "park" and "pleasuring ground."[18] Some legislators apparently realized that without intervention the land laws of the time would permit an ambitious or craven settler to claim lands like Yellowstone for private benefit. The inchoate sense among these legislators was that private ownership of such a place was grossly inappropriate—that places of such grandeur were best regarded as part of a larger American inheritance. Once that precedent had been set, it could be applied elsewhere—and it was applied, again, at Yosemite Valley in California, and then later at places like Wind Cave and Mount Rainier.

Yet these reservations, intended to preserve such sites as public, took no account of the fact that the same Native American communities which were everywhere threatened by settler activity were equally threatened by high-minded conservation activity. The notion of "reservation," applied to apparent wilderness, obscured that the reserved lands had, for thousands of years prior, been home to communities now severely diminished.

Conclusion

Recall the question that began this chapter. Why are *these* the public lands? And why are they located so disproportionately in the American West? Why are some biomes, such as mountain forests, so well represented in the federal land system, while rich grasslands are not? Why do land boundaries follow the cardinal directions of the compass rather than the physical features of the landscape?

The story of land acquisition and disposal goes a long way toward answering these questions. Congress had no plans for a comprehensive system of public lands in the nineteenth century. But federal politicians, bent on national expansion, drove a campaign of land acquisition, gaining sovereignty from other nations and then ruthlessly dispossessing Indigenous peoples of their land. Even as this process continued, the government began transferring land to settlers, states, and corporations in a process that was often chaotic and lawless. Settlers and prospectors claimed the best and richest lands for themselves while a distant Congress struggled to rationalize and regulate their claims. The public lands of today are, in large part, what was left behind.

But as we'll see, the insatiable demand of settlement caused others to fear the monopolization of western resources and the loss of frontier splendor. Slowly, federal policy shifted toward reservation, preservation, and conservation. Accordingly, today's public lands are a complex inheritance, a mix of unwanted leftovers and deliberately chosen national treasures. In the next chapter, we'll look more closely at how and why some of the federal public lands were reserved, and we'll begin to learn about the federal agencies that manage these lands on behalf of the American people.

Study Questions

1 How did lawmakers in Washington, D.C. interact with the messy processes of settlement that were playing out on the frontier?
2 What does the rectangular survey system reveal about the young nation's priorities regarding the use of public lands?
3 Compare and contrast the patterns of land disposition created by the process of statehood and the construction of transcontinental railroads.

Resources for Further Study

Banner, Stuart. *How the Indians Lost Their Land: Law and Power on the Frontier* (Cambridge, MA: Belknap/Harvard University Press, 2005).

Churchill, Ward. "The Black Hills Are Not for Sale: A Summary of the Lakota Struggle for the 1868 Treaty Territory." *Journal of Ethnic Studies* 18, no. 1 (1988): 127–142.

Gates, Paul Wallace. *History of Public Land Law Development* (Washington, D.C.: Arno Press, 1979).

Lee, Robert. "Accounting for Conquest." *The Journal of American History* 103, no. 3 (2017): 921–942.

Linklater, Andro. *Measuring America: How the United States Was Shaped by the Greatest Land Sale in History* (New York: Plume, 2002).

Notes

1 Mark David Spence, *Dispossessing the Wilderness: Indian Removal and the Making of the National Parks* (New York, NY: Oxford University Press, 1999), 4 (emphasis added).

2 See *Johnson v. M'Intosh*, 21 U.S. 543 (1823); Stuart Banner, *How the Indians Lost Their Land: Law and Power on the Frontier* (Cambridge, MA: Belknap/Harvard University Press, 2005).

3 In truth, the process was not nearly so straightforward, for the preceding European powers had made tens of thousands of land grants to their own citizens. The U.S. didn't always accept their claims of ownership at face value; in fact, one of the most labor-intensive tasks facing the early U.S. government was simply to adjudicate the entire body of private land claims—which totaled some 45 million acres. It took decades, multiple congressional committees, and frequent lawsuits to complete the process. See Paul Wallace Gates, *History of Public Land Law Development* (Washington, D.C.: Arno Press, 1979), 87–119; Gregory Ablavsky, "Getting Public Rights Wrong: The Lost History of Private Land Claims," *Stanford Law Review* 74 (2022): 277–351.

4 Felix Cohen, "Original Indian Title," *Minnesota Law Review* 32 (1947): 28–59, 35.

5 The so-called "Indian Commerce Clause" of the U.S. Constitution (article I, section 8, clause 3) authorized Congress to "regulate Commerce with foreign Nations, and among the several States, and with the Indian Tribes," and Congress used this power to enact the Indian Non-intercourse Act (enacted on July 22, 1790), which prohibited Indian land sales except by the authority of the United States. Please note that we use the term "Indian" advisedly throughout this text and with great respect for the people to whom that term refers. The word has become a technical term of art insofar as it is regularly employed in the United States Code and remains the term used in such contexts by tribal members and non-members alike.

6 Robert Lee, "Accounting for Conquest," *The Journal of American History* 103, no. 3 (2017): 921–942; Ward Churchill, "The Black Hills Are Not for Sale: A Summary of the Lakota Struggle for the 1868 Treaty Territory," *Journal of Ethnic Studies* 18, no. 1 (1988): 127–142.

7 In fact, "Indian tribes have treaty rights to half the salmon runs in the states of Washington and Oregon and claims to potentially vast quantities of water throughout the Western United States." James Rasband et al., *Natural Resources Law and Policy* (St. Paul, MN: Foundation Press, 3rd ed 2016), 99.

8 James Muhn and Hanson R. Stuart, *Opportunity and Challenge: The Story of the BLM* (Washington, D.C.: U.S. Government Printing Office, 1988), 16.

9 See Gates, *History of Public Land Law Development*, 145 ff. for a detailed account.

10 *Buford v. Houtz*, 133 U.S. 320, 326–327 (1890).

11 This remark is commonly attributed to the eighteenth-century English landscape gardener William Kent.

12 John B. Wright, "Land Tenure: The Spatial Musculature of the American West," in *Western Places, American Myths: How We Think About the West*, ed. Gary J. Hausladen (Reno, NV: University of Nevada Press, 2003), 92.

13 See Jon A. Souder and Sally K. Fairfax, *State Trust Lands: History, Management, and Sustainable Use* (Lawrence, KS: University Press of Kansas, 1996), at 17–36.

14 Benjamin Horace Hibbard, *A History of Public Land Policies* (Madison, WI: The University of Wisconsin Press, 1965), at 196.

15 Under the Mining Law, these rectangular claims are limited to 1,500 feet by 600 feet, or 20.7 acres.

16 *Leo Sheep Co. v. United States*, 440 U.S. 668 (1979); but for a contrary view, see *Iron Bar Holdings v. Cape*, 131 F.4th 1153 (10th Cir. 2025).

17 William Cronon, *Nature's Metropolis: Chicago and the Great West* (New York, NY: W.W. Norton, 1991).

18 Act establishing Yellowstone National Park, March 1, 1872.

3 From Retention to Management: Forming Today's Public Lands

Introduction

The previous chapter discussed the rough outlines of the public domain, explaining why the land has such strange boundaries. The chapter ended with the concept of reservation—the idea that federal lawmakers, toward the end of the 1800s, began to "reserve" certain land from the public domain, such as the Yellowstone area. This chapter picks up where the prior one left off. It will discuss how the process of reservation—a process which was neither straightforward nor uncontroversial—led to a jumble of categories of public land, which are often intermixed. It will explain why there are National Parks, National Forests, National Wildlife Refuges, National Monuments, and so on. Finally, it will start to describe how, in just one hundred years, the management of the public lands evolved from a bare-bones, no-budget operation to the modern, multi-agency bureaucracy that we have today.

The Jumble of Public Lands

Let's begin with another map. Take a look at Figure 3.1.

The various shades correspond to different categories of public lands in Utah. The green lands are National Forests. The small grey-blue areas are National Wildlife Refuges. The dark brown patches are Utah's five National Parks, as well as the Glen Canyon National Recreation Area, which is also managed by the National Park Service. Pale brown represents the Bureau of Land Management (BLM). The remaining uncolored areas are in private or state ownership—and as you can see, those uncolored areas are far less than half the land in Utah.

The maps of other western states would also look like a multi-colored jumble. How did this confusing mosaic come to be? Chapter One briefly described North America's ecological and physical characteristics; Chapter Two explained how the federal government acquired land, then promptly tried to dispose of it. The rest of this chapter discusses what the government did with the land it retained. But first, it will explain why Americans began to demand land retention in the first place.

Reservation and Withdrawal of the Public Lands

The United States changed a great deal during the 1800s. At the beginning of the century, the U.S. was a fledgling country along the Atlantic seaboard, less than a quarter of its present size. Western North America was claimed by Spain, France, and

DOI: 10.4324/9781003392514-3

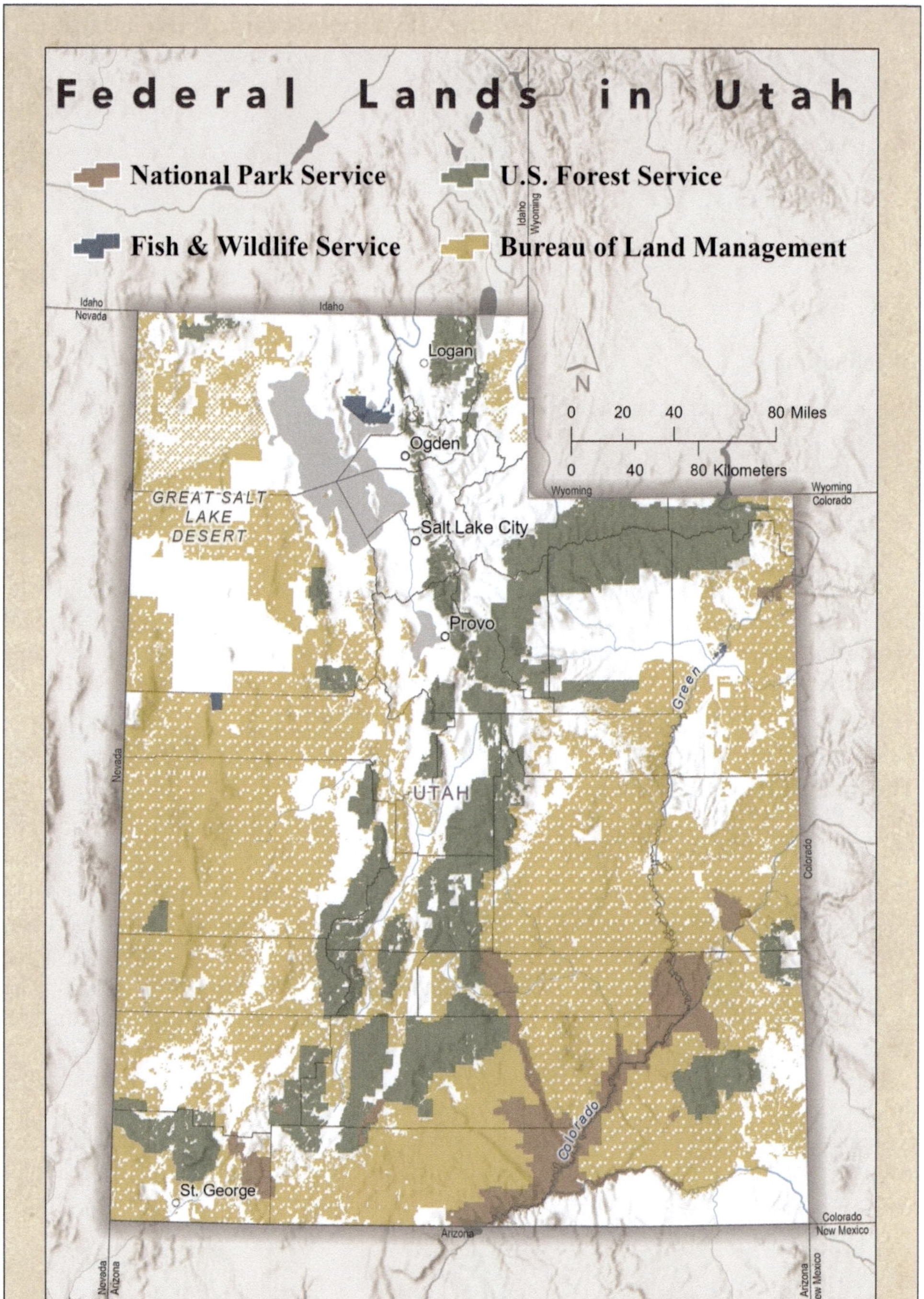

Figure 3.1 The state of Utah and its federal public lands. Map by Addison Arendt.

Britain but predominantly inhabited by Native Americans. By 1890, when the United States Census Bureau announced that the nation lacked any meaningful frontier territory (by which it meant land with fewer than two inhabitants per square mile), the country's current political boundaries were largely in place. Almost all of the land in the eastern half of the country had been transferred from federal ownership to state and private ownership.

By the turn of the century, two things were becoming clear about federal land policy. First, the land disposal laws—the laws that had transferred so much federal land to state and private ownership, as described in Chapter Two—were ineffective in the arid and mountainous West. Where forage was scarce, the disposal laws' standard 160-acre grant was grossly insufficient. This was explained to Congress by John Wesley Powell, who led the Geographical and Geological Survey of the Rocky Mountain Region in the late 1860s and 1870s, a campaign immortalized by his expedition down the Colorado River in 1869. In Powell's 1879 *Report on the Lands of the Arid Region of the United States*, the master surveyor explained to Congress that the East's pattern of agricultural settlement was doomed in the West.[1] He urged Congress to create a new policy encouraging irrigated agriculture, and he proposed a new grazing homestead law that would give ranchers 2,560 acres rather than 160. Congress temporized, but a more accurate picture of the distant West was taking shape in Washington, D.C.

Second, the negative impacts of settlement and economic expansion were becoming clear. Bison populations had plummeted close to extinction; eastern and Great Lakes forests had been felled; rivers had been diverted; grasslands were overgrazed by domestic livestock. Had aerial photography been available in 1890, one would have already seen the abstract map lines of the rectangular survey carved onto the prairie landscape in the form of roads and fields. The country's keenest observers recognized that something profound was changing; something was being lost. In 1864, George Perkins Marsh issued a jeremiad titled *Man and Nature*, a prescient warning about the unsustainability of American land and resource use. Marsh served as ambassador to Italy from 1861 until his death in 1882 and was a careful observer of the Mediterranean landscape. With a natural determinism characteristic of his day, Marsh argued that the Roman Empire had collapsed in part because the Romans and related cultures had violated the settled laws of nature:

> It appears, then, that the fairest and fruitfulest provinces of the Roman Empire, precisely that portion of terrestrial surface, in short, which, about the commencement of the Christian era, was endowed with the greatest superiority of soil, climate, and position, which had been carried to the highest pitch of physical improvement, and which thus combined the natural and artificial conditions best fitting it for the habitation and enjoyment of a dense and highly refined and cultivated population, is now completely exhausted of its fertility, or so diminished in productiveness, as, with the exception of a few favored oases that have escaped the general ruin, to be no longer capable of affording sustenance to civilized man.[2]

Unless Americans changed their practices, he warned, their nation would destroy its natural heritage and collapse like the Roman Empire.

After the Civil War, public concern about a potential timber famine and other resource shortages, along with growing interest in scenic preservation, drove a new conservation movement. In numerous books and articles, the leaders of this movement described the changes in the West—but more importantly, they took their message directly to political leaders in Washington, D.C., who used what authority they had to staunch the bleeding. The tools most readily at their disposal were "reservation" and "withdrawal." These two legal terms-of-art were roughly synonymous: "withdrawal" referred to exempting land from the operation of the Homestead Act and other land disposal laws, whereas "reservation" referred to setting lands aside for particular purposes (see Box 3.1). The nineteenth-century understanding of the purpose of public lands—that they should be made available for settlement as rapidly as possible—was giving way to a new understanding which emphasized conservation rather than settlement.

BOX 3.1 Reservation vs. Withdrawal

Both reservations and withdrawals still happen today, and the distinction between them, though subtle, matters to land managers. If you follow the news about the public lands, you'll sometimes hear that a president or federal agency has *withdrawn* land from the operation of a law that would otherwise allow private parties to conduct an activity on that land (for an example, flip ahead to Figure 13.4). Usually, the issue is mining. Presidents can use their authority to withdraw certain land from mining laws that otherwise would allow mining companies to prospect and extract minerals.

In other instances, you'll hear that the government has created a new *reservation* of some kind (such as a new wildlife refuge) or perhaps added land to an existing one. The reservation will generally have a specific purpose and will be accompanied by new legal instructions to the managing agency. Often the reservation includes a withdrawal—the reserved land is also withdrawn from the general laws that would allow, say, mining or grazing.

The First National Parks

An early manifestation of this new understanding, which gradually took hold in the late 1800s, was a growing sense that settlers should not be allowed to homestead on lands of particular scenic or natural merit. The first debates along these lines addressed Yosemite Valley in California and the Yellowstone area of Wyoming. Early visitors to these sites feared that profiteers could claim land under the guise of settlement and effectively monopolize access. Congress briefly discussed these concerns, and its initial response was to simply pass the problem off to the host state by granting it the relevant land. Congress gave Yosemite Valley to the State of California in 1864 without debate and without any real suggestion that the federal government could itself reserve the land from claimants. But when explorations of Yellowstone in 1869 and 1870 led to similar calls for a congressional reservation, the give-it-to-the-state option was unavailable because Wyoming was not yet a state. If settlement was to be taken off the table, Congress itself would have to amend federal land law as to this specific territory.

It is difficult now to appreciate why restricting settlement in Yellowstone was controversial, but it was—in fact, it was far more controversial than granting Yosemite to California. The laws that authorized settlement had been suspended previously for small bits of land containing springs, forests, or minerals of interest to one legislator or another. But now came a bill which purported to exclude from settlement a rectangle of 2 million acres, larger than Delaware or Rhode Island. The debate in Congress was fierce. Settlement of the West was understood as a God-given right, the manifest destiny of the American people. Without settlement, territories could not apply for statement. Settlement was necessary for national defense and economic growth. So a federal limitation on homesteading was perceived as a threat not only to specific locations but to western states and territories generally, which depended on population growth to attain statehood and on resource extraction for economic survival. For decades to come, withdrawal of public domain land from the operation of the land disposal statutes was regarded as an act of hostility toward the emerging Western establishment.

But on the other side of the ledger there were the magical and mysterious geysers, the roiling waterfalls, the gaping canyons of the Yellowstone River. Railroads recognized immediately the area's tourist potential. Some worried that, under the law as it then stood, a lucky claimant could control access to such wonders on whatever terms he chose. In the words of one Senator,

> Here is a region of country away up in the Rocky Mountains, where there are the most wonderful geysers on the face of the earth; a country that is not likely ever to be inhabited for the purposes of agriculture; but it is possible that some person may go there and plant himself right across the only path that leads to these wonders, and charge every man that passes along between the gorges of these mountains a fee of a dollar or five dollars.[3]

What did America need most: protection *for* the settler-as-entrepreneur? Or protection *from* the settler-as-monopolist?

The Yellowstone Park Act passed Congress and was signed into law on March 1, 1872. The Act declared the area to be

> reserved and withdrawn from settlement, occupancy, or sale under the laws of the United States, and dedicated and set apart as a public park or pleasuring-ground for the benefit and enjoyment of the people; and all persons who shall locate or settle upon or occupy the same, or any part thereof, except as hereinafter provided, shall be considered trespassers and removed therefrom.

Just as importantly, the legislation placed the park "under the exclusive control of the Secretary of the Interior" and gave the Secretary the authority to issue rules and regulations to protect it.[4]

The Yellowstone legislation provided a template, but it also began an experiment. In some ways, Yellowstone and Yosemite represented two competing modes of protection. Yellowstone was now off-limits to claimants and ostensibly under federal control; Yosemite was entrusted to the care of the state of California.

The Yellowstone model prevailed. In the years to come, California state officials seemed unable to fend off the private profiteers who wished to exploit Yosemite's grandeur for commercial gain. Yellowstone presented challenges for the Interior Department, but none had to do with ownership per se, and most could be solved with small appropriations from Congress. So a few years later, when Michigan legislators sought protection for Mackinac Island in 1875, they proposed a national park rather than a state park.[5] And partly due to California's mishaps, Congress in 1890 created three national parks in California: first Yosemite, but also Sequoia National Park and

General Grant National Park (now Kings Canyon). Then followed Mount Rainier in Washington State in 1899 and Crater Lake in Oregon in 1902. The idea of the national park was becoming firmly established.

The Forest Reserves

National parks were just the tip of the iceberg. In the late 1800s, timber theft and over-harvesting also became matters of national concern. Demand for timber was incessant. Enormous volumes of timber were necessary for railroad ties, for houses, for fuel, for building mines, and for erecting bridges and trestles, among other things. And where forests were harvested, erosion filled rivers and streams with soil and silt, degrading water quality.

In 1891, President William Henry Harrison signed legislation that had been brewing in Congress for years. The bill, called the General Revision Act,[6] attracted little public attention but was transformational for the public lands. The Act's final section, Section 24, authorized the President to "set apart and reserve … public lands wholly or in part covered with timber or undergrowth, whether of commercial value or not, as public reservations … ."

Not Congress but the *President* was given the authority to create forest reserves. This was an important innovation. Congress tended to bicker, often taking years to arrive at compromise. Members of Congress were beholden to local constituencies. But the President had a single, national constituency and could act with alacrity. This political dynamic mattered a great deal because, as noted above, restrictions on settlement were deeply unpopular in the western states. Taking the wrong side on this issue could doom a local congressional representative. The President, however, was more insulated from regional political forces.

But not completely insulated. President Harrison and the two presidents after him, President Grover Cleveland and Teddy Roosevelt, together "set apart" over 170 million acres of public lands as "forest reserves." The western states couldn't stop them directly, but they could marshal opposition in Congress—which they did, successfully, during Roosevelt's presidency, amending the law to disallow presidents from creating any more forest reserves without congressional authorization. It was a bit too late; by that time, the presidents had assembled the lands that form the heart of today's National Forest system.

FIFTY-FIRST CONGRESS. SESS. II. CHS. 561–563. 1891.

law complied with as to settlement, said subsequent entries be, and the same are hereby, confirmed.

SEC. 24 That the President of the United States may, from time to time, set apart and reserve, in any State or Territory having public land bearing forests, in any part of the public lands wholly or in part covered with timber or undergrowth, whether of commercial value or not, as public reservations, and the President shall, by public proclamation, declare the establishment of such reservations and the limits thereof.

Approved, March 3, 1891.

Figure 3.2 An early printing of Section 24 of the General Revision Act of 1891. Source: U.S. Congress.

BOX 3.2 "The Midnight Reserves"

Western legislators were deeply opposed to the President's unilateral authority to create forest reserves. By 1907, they succeeded in persuading Congress to strip the President of this authority. Fearing a veto from President Roosevelt, Congress cleverly joined the forest reserve legislation to other provisions that Roosevelt desperately wanted. But Roosevelt was unfazed by this tactic. Rather than veto the law, he spent a hasty few days creating one last, massive batch of 31 new forest reserves, covering tens of millions of acres, before the new law took effect. These epic reserves came to be nicknamed "The Midnight Reserves" in honor of their last-minute creation, and they helped earn Teddy Roosevelt his reputation as one of our most conservation-minded presidents.

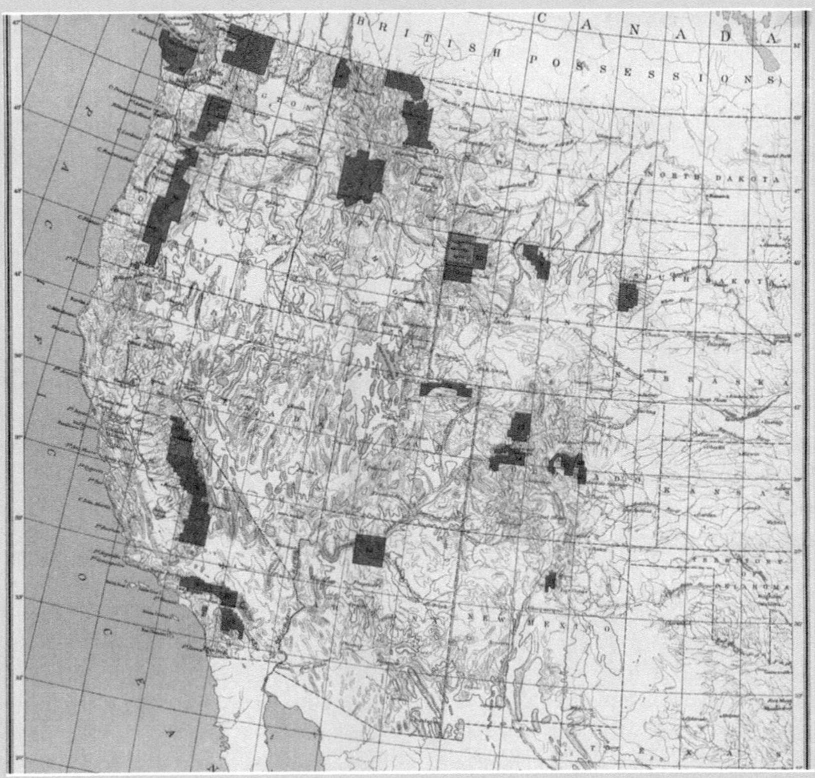

Figure 3.3 The extent of the U.S. forest reserves in 1898 (Figure 3.3) and in 1908 (Figure 3.4). The expansion of the reserves was attributable almost entirely to President Teddy Roosevelt. Source: U.S. Geological Survey; U.S. Forest Service.

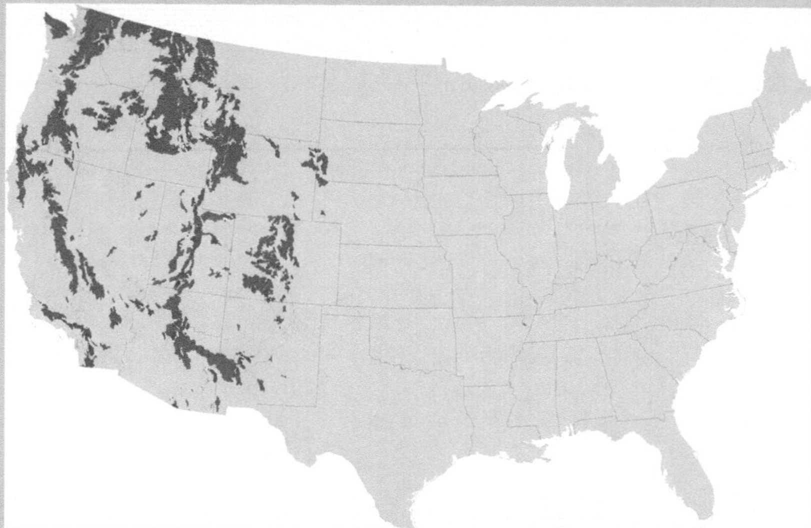

Figure 3.4 The extent of the U.S. forest reserves in 1898 (Figure 3.3) and in 1908 (Figure 3.4). The expansion of the reserves was attributable almost entirely to President Teddy Roosevelt. Source: U.S. Geological Survey; U.S. Forest Service.

Wildlife Refuges

President Roosevelt's influence on the formation of today's public lands was arguably larger than that of any other president. In addition to his massive additions to the forest reserves, Roosevelt also created the nation's first wildlife refuges.[7] The initial purpose of these refuges, and one that is still central to the National Wildlife Refuge System, was the protection of bird populations. A longtime lover of birds, Roosevelt was disturbed to learn in the early 1900s that the commercial market for exotic feathers—used for women's hats and other fashion accessories—was threatening certain bird populations around the U.S. and elsewhere. His first reserve, created in 1903, protected the stunning brown pelicans that regularly occupied Pelican Island in Southeast Florida.

Roosevelt went on to designate over 50 wildlife refuges around the country, nearly all of which were located on flyways for various migratory birds. But unlike the forest reserves, these refuges were created without any explicit authorization from Congress. Could the President simply set aside public land without any law giving him permission to do so? Roosevelt believed that he could. He argued that, by virtue of his office, he served as the manager or proprietor of the public lands, and that he could remove them from sale just like a store clerk could pull an item off a shelf and withhold it from sale. In 1906, Congress ratified Roosevelt's wildlife refuge reservations, stamping its approval on this uncommon use of presidential power. This episode remains highly significant today because of the precedent it established regarding presidential authority over the public lands. As you keep reading, be sure to distinguish between actions of Congress and actions of the President or the officials and agencies of the executive branch. Many, many lawsuits against the federal land agencies involve the question of whether those agencies have proper legal authorization for their actions.

BOX 3.3 The Reservation of Pelican Island

The following excerpt, taken from Douglas Brinkley's excellent biography of Teddy Roosevelt, *Wilderness Warrior*, describes some of the difficulties in setting land aside for conservation. Note the references to the land survey, to the legal preference for homesteaders, and to the propriety of unilateral presidential action.

> The American Ornithologists Union had been trying to purchase Pelican Island outright from the federal government for three years, to no avail … For years AOU had demanded that Pelican Island be surveyed—a prerequisite for placing a purchase bid on it. Now, with the official 1902 survey about to be filed, AOU felt boxed in. Legally, homesteaders' applications had to be given preference when [General Land Office] land was sold. With homestead filings imminent, the AOU's application would be shunned or given a low priority. And that meant the brown pelicans might not survive as a species on the Atlantic coast.
>
> A hunter and conservationist himself, [Land Office Commissioner William] Richards wanted to help AOU, and he summoned Charles L. DuBois, his chief of the Public Surveys Division, into the meeting. Was there an ingenious way to circumvent the homesteaders-first provision? DuBois, a jurist who always dotted the i's and crossed the t's, at first said no. But he offered [the AOU] one long-shot alternative. President Roosevelt could make Pelican Island a bird refuge by issuing an Executive Order. Worried that a firestorm would ensue if the U.S. Department of the Interior seemed to be in collusion with the AOU, DuBois instead suggested pushing the Executive Order through the USDA, where it would go virtually unnoticed …
>
> Now that the AOU had a credible, legal way to protect Pelican Island, [the AOU] wrote to Secretary of Agriculture James Wilson asking that a federal bird reservation be created … [The AOU then] pushed for a meeting with the president about Pelican Island. Time was of the essence. With minimal difficulty [they] procured a White House meeting that March.
>
> After listening attentively to their description of Pelican Island's quandary, and sickened by the update on the plumers' slaughter for millinery ornaments, Roosevelt asked, "Is there any law that will prevent me from declaring Pelican Island a Federal Bird Reservation?" The answer was a decided "No"; the island, after all, was federal property. "Very well then," Roosevelt said with marvelous quickness. "I So Declare It."[8]

The Antiquities Act and National Monuments

The issue of presidential (as opposed to congressional) authorization for withdrawals and reservations of public lands was a persistent one. It was as though a conversation was occurring between not only the particular officeholders of the time but between the very branches of government themselves. The key driver of this debate, almost always, was some new public awareness about a lamentable depredation of the public's land. First it was the loss of forestland, then of bird habitat. Shortly thereafter, Congress focused its attention on another problem: invaluable artifacts, such as petroglyphs or shards of ancient pottery, were being stolen from or destroyed on public lands with alarming frequency. Looters were walking off with tens of thousands of items and selling them for profit. The situation demanded a rapid government response, but Congress could not rapidly identify the lands from which artifacts were being taken. The President or an executive branch agency could do that sort of work, but to act unilaterally, as Roosevelt had in the case of Pelican Island, was legally suspect. And unauthorized presidential reservations came at the expense of Congress's prerogative to make public lands policy.

So in response to the problem of stolen or defaced cultural artifacts, Congress in 1906 passed legislation called the Antiquities Act. The law authorized the President to reserve, unilaterally, "national monuments" containing "objects of historic and scientific interest." To appease westerners, who feared that these withdrawals would swell to the size and scale of the forest reserves, the Act limited the President's power to reserve land to "the smallest area compatible with proper care and management of the objects to be protected."[9]

In hindsight, it was naïve of Congress to believe that this limitation would hold. Presidential power is like a one-way ratchet. When presidents are faced with challenge or opportunity—as we saw with Thomas Jefferson and the Louisiana Purchase—they tend to test the limits of their powers. If they succeed (and they often do), a precedent is established, and subsequent presidents generally find such powers useful for their own purposes. So it was with the Antiquities Act. As Hal Rothman has written:

> [The Act] became the primary federal tool in urgent situations, and beginning with Theodore Roosevelt, it was used in ways that closely resembled uses of the Forest Reserve Act. Areas far larger than ever conceived were reserved under the auspices of the Act … [It] became the way to protect public land in the United States quickly and effectively.[10]

The first several monuments reserved under the Act were small, such as Devil's Tower in northeastern Wyoming (see Figure 3.5). But they rapidly became larger and larger. Before Roosevelt's term ended, he created the Grand Canyon National Monument, a behemoth of over 800,000 acres. Many Arizonans protested, including Arizona Senator Ralph Cameron, who purported to own several mining claims conveniently located at some of the Canyon's best viewpoints. Cameron argued that the President had exceeded his authority under the Act, but the U.S. Supreme Court, which ultimately heard his case, disagreed. The Court upheld Roosevelt's designation, acknowledging the Canyon's extraordinary scientific interest. The "smallest area compatible" limitation was of little consequence when the "objects to be protected" were of such an extraordinary scale.[11]

The power conferred by the Antiquities Act has proven to be one of the most useful conservation instruments in any president's arsenal. It has been used over one hundred times, and several patterns have emerged. First, national monument status is often a stepping stone, in that many monuments are later "upgraded" to national parks. Second, even to this day, presidents often use monument designations to burnish their environmental credentials at the end of their terms of office, creating sizeable monuments in their final weeks in the White House.

Figure 3.5 Devil's Tower, Wyoming, site of the first National Monument. Source: National Park Service.

Coal, Oil, and Minerals

Thus far, we've discussed the origins of land categories that are quite familiar: National Parks, National Forests, National Wildlife Refuges, and National Monuments. But a great deal of land remains in public ownership not because it was placed in one of these special categories but because it held valuable mineral resources. Coal, oil, and other minerals abound on the western public lands. At roughly the same time that federal officials were beginning to protect scenic, ecological, or cultural resources, they also recognized the need to set aside mineral-rich lands. But the reasons had little to do with conservation, at least as we understand that word today.[12]

In Chapter Two, you learned about the General Mining Act of 1872 and how it essentially gave land away for free to successful prospectors. Massive quantities of gold and silver, removed from land technically owned by the federal government, were making thousands of people enormously wealthy. The same thing was happening across the West with respect to other, more common minerals, such as coal and oil. Eventually, easterners began to object. They argued that the mineral wealth on Western land belonged to all Americans, not just to prospectors, coal miners, and the like. It seemed foolish to give away resources that the government would then have to buy, such as the oil used to power navy ships. The question was, what system would best distribute mineral wealth for the benefit of the general public?

Take coal, for example. Throughout the 1800s, coal was an enormously important fuel. But the laws that disposed of the public domain were oriented toward settlers and were ill-suited to deal with large, industrial-scale coal deposits. Genuine settlers didn't have the ability to mine coal, and genuine coal miners didn't qualify to purchase public domain land under laws like the Homestead Act. To squeeze their claims into these laws, coal mining entities often had to lie about their intentions or bribe individuals to file homestead claims on their behalf. As critics began to allege that coal firms were monopolizing a resource that rightfully belonged to the public, Congress first tried to address the problem by increasing the price of coal lands—up to $20 per acre in the late 1800s, and perhaps four or five times that amount in the early 1900s. But this was a drop in the bucket, given that an acre of coal land could produce many thousands of dollars' worth of coal. The government was groping for a system that would encourage coal development but give a fair return to the American public.

In the early 1900s, the Roosevelt Administration seemed to hit upon a solution. Instead of *selling* coal- or oil-rich lands, they could be *leased* to mining entities, who would be obligated to pay a royalty to the government, equal to some fraction (say, one-eighth) of the value of the fuel produced. Congress took years to be persuaded of the wisdom of this arrangement; it wasn't until 1920 that the Mineral Leasing Act was passed. In the meantime, Roosevelt used his executive authority to simply withdraw coal or oil lands, just as he'd reserved wildlife and forest reserves and national monuments.

Congress willingly ratified Roosevelt's withdrawals and also hit upon a new tactic. Reluctant to simply withdraw mineral land that was also agriculturally productive, Congress instead chose to sever the land's surface from the minerals underneath. The surface would still be open to homesteaders and other agricultural claimants. The mineral rights beneath, however, would belong to the public until they could be sold or properly disposed. To this day, there are millions of acres of "split estate" land—land whose surface is owned by one party but whose minerals or subsurface rights are owned by another.

Lands overlying oil and gas fields, as well as those containing fertilizer minerals like phosphate, were treated in a similar fashion to coal lands. Perhaps surprisingly, so-called "hardrock minerals"—gold, silver, copper, and the like—remain under the "free mining" policy of the Mining Law to this day. There will be much more to say about this later in the book, but for now the upshot is that some of today's public lands—especially land managed by the BLM—would today be in private ownership had it not been set aside due to its mineral potential.

Building the Land Management Agencies

As you've now seen, much of our public land was set aside land from settlement or resource exploitation in order to protect it. Setting it aside was only half the battle. The other half was creating organizations capable of managing and caring for the land appropriately. Today we take for granted that government can oversee enormous projects, from Social Security to the Space Shuttle. Indeed, there seems to be an agency for everything. But in the early 1900s, the federal government was still quite small. To manage the emerging public lands required not merely a new set of agencies but a new vision of what the federal government could be. The following pages will discuss the emergence of this new vision by providing a brief account of the origins of the principal federal land management agencies. Chapters Five through Eight will provide much more detail about how those agencies function today.

The National Park Service

When Yellowstone was first designated a National Park, the Interior Secretary was given authority to make rules governing the park. But who would enforce the rules? Interior had no staff in Wyoming (which wasn't even a state when Yellowstone was created) and no permanent office of land management. The Yellowstone legislation appointed a single superintendent, but he didn't even reside at the park, and he received no appropriations of any kind, not even a salary, for 5 years. Eventually the Army—one of the few well-staffed organizations in the entire federal system—was called upon to oversee the Park, to prevent vandalism and illegal settlement, and to generally maintain order. As more parks were created, the same problem arose elsewhere. The following description gives some sense of the disarray:

> The division of authority among the parks, and even inside a single park, came close to chaos. In Yellowstone all improvements and their appropriations were managed by an officer of the Army Corps of Engineers, who answered to neither the Interior Department nor the park superintendent; the Superintendent was himself an army officer, appointed by the Secretary of War; and "exclusive control" rested with the Secretary of the Interior. Crater Lake and Mount Rainier, like Yellowstone, used Army engineers for road-building and improvements, but in those two parks the superintendent was a civilian, appointed by the Secretary of the Interior.[13]

Eventually, calls arose for a centralized office or bureau to manage the parks, and Congress passed legislation to that effect in 1916. The first director of the National Park Service (NPS) was Stephen Mather, a wealthy businessman. His public stature and independent wealth made him a powerful leader, able to build an organization from the ground up, to confront opponents in Congress, and to use his own personal funds to finance initiatives that Congress refused.

Mather's tenure as director lasted only about 10 years, but during that time he provided the NPS with a firm foundation. As an industrialist, Mather's marketing experience shaped his approach to agency management. He realized that he needed a powerful constituency to support the parks, so he expended vast energies to make the parks accessible to the general public. He brokered deals with hotels, railways, and other concessioners; he promoted the parks with advertising and other efforts at publicity; and he built facilities for visitors, both for purposes of accommodation and of interpretation of park resources. By midcentury, the NPS was a thriving, well-respected agency, and tourists were flocking to the national parks in extraordinary numbers.

The Forest Service

Like the NPS, the U.S. Forest Service (USFS) had a powerful early leader, Gifford Pinchot, whose influence on the Service lingered for decades. When the first forest reserves were created in the 1890s, responsibility for their oversight lay within the Interior Department. There were few professional foresters in Interior; however, in fact, the idea of "professional" forestry barely existed in the United States. In Europe, by contrast, trees had been grown as crops for hundreds of years; there were schools of forestry (which also did not exist in the U.S.); and the methods of forestry were far

Figure 3.6 Stephen Mather, the first director of the National Park Service. Source: National Park Service.

more sophisticated. After all, timber could be understood as an agricultural product like any other. Pinchot, who had studied forestry in both France and Germany, couldn't find a receptive ear in Interior for his ideas about forest management. In 1898, he was named chief of the Division of Forestry within the Department of Agriculture, where he managed no land but had a staff of several dozen scientists who provided technical assistance for both private and public forest management. Most importantly, Pinchot befriended Teddy Roosevelt, who in 1905 placed the forest reserves under Pinchot's care and renamed his organization the Forest Service.

Under Pinchot, the Forest Service quickly became one of the most professionalized and respected federal agencies. Pinchot's foresters were dedicated to producing timber efficiently while also ensuring that high-quality timberland was reseeded. True to the ethos of the Agriculture Department, Pinchot treated timber as a crop and sought to achieve the maximum sustainable yield from the forests in his care. Output from the National Forests (as the forest reserves came to be called) grew impressively, reaching an annual harvest of 10 billion board feet in the 1960s, a level that would be sustained until the early 1990s. By that time, the USFS's reputation as a cadre of professional foresters dedicated to "getting out the [timber] cut" became anachronistic in an era of growing concern about habitat loss and biodiversity. But more about that in Chapter Seven.

Figure 3.7 President Teddy Roosevelt with Gifford Pinchot on the Mississippi River in 1907. Source: Library of Congress.

The Grazing Service

The National Park Service and the U.S. Forest Service have enjoyed sterling reputations for much of their existence. Not all land management agencies, however, have benefited from such goodwill. The story of the U.S. Grazing Service, a precursor to the Bureau of Land Management, helps convey a more realistic portrayal of the range of agency competence and efficacy.

One of the most incendiary issues in the management of the public lands in the early 1900s had to do with livestock grazing. Cattle ranchers and sheepherders, after claiming as much land as the law would allow, made a practice of allowing their massive herds to roam across hundreds of miles of the open public range. The fabled cattle drives of western lore date to this period. In some instances, the consequences of overgrazing were disastrous: forage was eaten so bare that it could not regenerate, causing erosion, soil loss, stream pollution, and cattle die-offs.

Although the Forest Service regulated grazing within the National Forests, the vast unreserved portions of the western range did not come under general grazing regulation until the passage of the Taylor Grazing Act in 1934. Portions of the Act had been wending through Congress for years, but the bill did not gain passage until lawmakers were confronted with the infamous Dust Bowl episode, which testified dramatically to the importance of rangeland health. The Act created the Division of Grazing in the General Land Office, which became the independent U.S. Grazing Service in 1939. The Service was tasked with dividing the range into grazing districts; each district would calculate the sustainable carrying capacity of the local forage and allocate it to local ranchers, giving a preference to those who had historically grazed within that district.

The Grazing Service was nothing like the Forest Service. The latter was centralized and professional, full of expert foresters and horticulturalists. The Grazing Service had little in the way of a central organization. The real power lay in local grazing advisory boards, comprised mostly of local ranchers. If it sounds like the inmates were running the asylum, that's about right—the regulated community was, to some extent at least, tasked with creating its own rules and then monitoring their enforcement. Perhaps unsurprisingly, overgrazing continued, mitigated only slightly by operation of the Grazing Service.

Meanwhile, the General Land Office continued its work surveying the public domain, selling and granting public domain lands, and leasing and patenting mineral resources. It also leased grazing land that the Grazing Service had not yet organized into grazing districts. Shortly after World War II, the Grazing Service was combined with the General Land Office to form the Bureau of Land Management (BLM). This reorganization had little immediate effect on land management. Only with the passage of the Federal Land Management and Policy Act of 1976 (FLPMA; insiders pronounce it "flip-muh") did Congress finally declare that public lands managed by the BLM would remain in permanent federal ownership. FLPMA also empowered the BLM to build a professional range management program, which—at long last—enabled the agency to exert more independent power over grazing.

The Fish and Wildlife Service

What entity managed wildlife refuges, like the one at Pelican Island created by Teddy Roosevelt? Until 1940, it was the Bureau of Biological Survey (BBS), housed within the Department of Agriculture alongside the Forest Service. The BBS began in the 1880s as a tiny, three-person unit created by Congress to study bird migrations, especially insofar as large migrations affected agriculture. The enduring refuge emphasis on birds made the BBS a good fit for management of the early refuges. In 1940 the Bureau was joined with the Bureau of Fisheries to create the Fish and Wildlife Service (FWS) within the Department of the Interior. At roughly the same time, Congress began to increase funding for the acquisition of land useful for bird sanctuaries. Thus, although much of the land in today's wildlife refuges was reserved out of the unsettled portions of the public domain, a sizeable proportion was purchased by the federal government from private landowners. We will have much more to say about the FWS in Chapter Six.

Conclusion

By 1946, four federal land systems had taken shape, each with a distinct set of legal mandates and constituents.[14] Yet federal public land management looked very different than it does today. Federal law and agency expertise was directed almost exclusively to the management of individual resources. The NPS focused on managing scenery and wildlife as a resource for recreational visitors; the USFS focused on managing forests for timber production and water quality; the FWS focused on maintaining viable populations of desirable species; and the BLM carried on the priorities of its predecessors.

This mode of resource-focused land management allowed some of the agencies to foster friendly constituencies that supported the agencies in Congress. The NPS enjoyed the backing of auto tourists who came to the national parks in ever-larger numbers, particularly after World War II; the Forest Service received broad endorsement from the timber industry, which had grown increasingly dependent on national forests; and the Fish and Wildlife Service quickly became a favorite of hunters and anglers. Only the new BLM lacked a clear and supportive constituency.

But there were rumblings of discontent, even from within the agencies, about the overall trajectory of federal public land management. For one thing, the NPS and USFS could not keep up with the rapid expansion of outdoor recreation in the postwar period. But more broadly, tensions were mounting over the balance between land protection and development. That, coupled with a growing ecological awareness, put pressure on Congress and the agencies to develop a more comprehensive framework for federal public land management. The development of such a framework is the subject of this book's next chapter.

But before we move on, let's pause to make a few observations. Our system of public lands is not entirely coherent. The sharp distinctions visible on the colored map at the beginning of the chapter fade away when one looks at the land itself. The distinctions between National Parks, Forests, Wildlife Refuges, and other land categories owe largely to the twists and turns of history—to aggressive presidents trying to outmaneuver congressional resistance, to skirmishes between emergent conservationists and proponents of settlement and development. And in all cases, the land that was up for debate was only what remained after the feverish land rushes of the 1800s. At the table thus prepared, federal lawmakers and presidents could bicker amongst themselves over who got what. Only in 1976, with the passage of FLPMA (mentioned briefly above), did Congress make clear what many Americans could already discern—that the era of land disposal was over and that what the public then owned would remain the public lands in perpetuity.

This chapter has only provided the barest outline of processes that played out over a century. And it has focused only on political elites. For any particular unit of public land, one could excavate a wealth of history, uncovering the stories of countless individuals and their relationships with the land beneath them. To be sure, the lines on a modern map obscure much. But they also help guide us into the past and its people. You should by now be able to examine a map or a landscape and envision the forces that led to the current ownership and managerial arrangements there. You are well positioned to dive more deeply into specific stories and to situate individual accounts within those broader forces. Such forces were often chaotic, disorganized, and disruptive—or, in other words, democratic. And the story of democracy is the story of people. In the next chapter, we turn to examine how the American people have chosen to govern their lands through law.

Study Questions

1. Why did federal policy shift from a focus on disposing of public lands to one of retention and reservation toward the end of the nineteenth century?
2. Why was the withdrawal of public domain land from settlement so controversial among the Western establishment?
3. Compare the role of Congress and the President in setting aside public lands for particular purposes, such as Forest Reserves, Wildlife Refuges, and National Monuments.

Resources For Further Study

Brinkley, Douglas. *The Wilderness Warrior: Theodore Roosevelt and the Crusade for America* (New York, NY: Harper Collins, 2009).

Leshy, John D. *Our Common Ground: A History of America's Public Lands* (New Haven, CT: Yale University Press, 2021).

Pinchot, Gifford. *Breaking New Ground* (Washington, D.C.: Island Press, 1998).

Squillace, Mark. "The Monumental Legacy of the Antiquities Act of 1906." *Georgia Law Review* 37 (2003): 473–610.

Stegner, Wallace. *Beyond the Hundredth Meridian: John Wesley Powell and the Second Opening of the West* (New York, NY: Penguin, 1992 [1954]).

Notes

1 You can read more about Powell and his ideas in a classic book by Wallace Stegner, *Beyond the Hundredth Meridian: John Wesley Powell and the Second Opening of the West* (New York, NY: Penguin, 1992 [1954]).

2 George Perkins Marsh, *Man and Nature*, ed. David Lowenthal (Seattle, WA: University of Washington Press, 2003 [1864]), at 10.

3 Comment of Senator Trumbull of Illinois, *Congressional Globe*, Jan. 30, 1872, p. 697, quoted in John Ise, *Our National Park Policy: A Critical History* (Baltimore, MD: The Johns Hopkins Press, 1961), 16–17.

4 Act establishing Yellowstone National Park, March 1, 1872.

5 Ironically, the Mackinac Island property was subsequently deeded to the State of Michigan and is no longer in federal ownership.

6 The name of the General Revision Act referred to the Act's overarching purpose, which was to revise and update the nation's public land laws. A decade prior, Congress had appointed an expert commission to review the operation of those laws and to recommend improvements. It took over 10 years for Congress to sort through those recommendations and prepare legislation for President Harrison's signature.

7 Presidents Harrison and McKinley, who preceded Roosevelt, both used the 1891 General Revision Act (discussed earlier) to set aside as forest reserves some lands that are *now* designated as National Wildlife Refuges—but at the time, of course, that designation did not yet exist. See Robert L. Fischman, *The National Wildlife Refuges: Coordinating a Conservation System Through Law* (Washington, D.C.: Island Press, 2003), 34.

8 Douglas Brinkley, *The Wilderness Warrior: Theodore Roosevelt and the Crusade for America* (New York, NY: Harper Collins, 2009), 13–14.

9 Antiquities Act of 1906, Section 2, 16 U.S.C. § 432.

10 Hal Rothman, *America's National Monuments: The Politics of Preservation* (Lawrence, KS: The University Press of Kansas, 1989), 48. For an excellent legal history of the Antiquities Act, see Mark Squillace, "The Monumental Legacy of the Antiquities Act of 1906," *Georgia Law Review* 37 (2003): 473–610.

11 *Cameron v. United States*, 252 U.S. 450 (1920). See also Rothman, *America's National Monuments*, 69 ff.

12 Today, conservation means something like "preservation." To conserve something is to *save* it rather than *use* it. But a century ago, conservation referred equally to the wise use of a resource. Conservation was opposed to waste rather than to use. See Samuel P. Hays, *Conservation and the Gospel of Efficiency: The Progressive Conservation Movement, 1890–1920* (Pittsburgh, PA: University of Pittsburgh Press, 1999 [1959]).

13 Robert Shankland, *Steve Mather of the National Parks* (New York, NY: Knopf, 1951), 53 (quoted in Ise, *Our National Park Policy*, 186).

14 To be clear, the Bureau of Land Management was still implementing thousands of land disposal laws that applied to the remaining public domain, so it was not managing a designated land system in the same sense as the other three agencies.

4 Modern Public Land and Resource Law

Introduction

Most landowners manage their land according to their own ideas and preferences, within the broad boundaries of the law. For federal land managers, those boundaries are much narrower. To be sure, the law gives those managers some wiggle room, but they can't simply do what they please. Public lands law is a mixture of strict boundaries and flexible guidelines. Many rules are specific to particular systems, like the national parks or national forests. But others are general and apply to all the federal land agencies. This chapter provides an overview of public lands law. It will focus especially on several important statutes enacted in the 1960s and 70s. These laws—the Wilderness Act of 1964, the National Environmental Policy Act of 1968, and the Endangered Species Act of 1973—are ubiquitous in land management discussions today. It's difficult to overstate their impact, and by the end of this chapter, you should have a basic sense of what these laws require.

The Environmental Movement and Public Land Law

In the years after World War II, a growing number of Americans and their representatives in Congress grew frustrated with the patchwork of laws and resource programs that had accumulated in the nineteenth and early twentieth centuries. As the previous chapter made clear, Congress's primary concerns during that era were transferring land to private ownership, developing natural resources, preserving a few areas of exceptional scenery, and maintaining flows of renewable resources like timber and water. But as the modern environmental movement took shape during the 1960s, its leaders demanded a comprehensive, national environmental policy that would press all four federal land agencies to manage land on the basis of ecosystem health. The movement pitted itself against extractive industries (such as mining and timber harvesting) and the western state governments that depended on federal resources.

BOX 4.1 The Public Land Law Review Commission

In 1964, Congress created the Public Land Law Review Commission and tasked it with a comprehensive review of federal public land policy. Chaired by representative Wayne Aspinall of Colorado, the Commission spent the next 6 years gathering data, holding hearings, and publishing studies. It produced two enduring works, both of which remain standard references.

DOI: 10.4324/9781003392514-4

The first is a volume titled *History of Public Land Law Development*. As its name suggests, it is a comprehensive account of public land law; it was prepared for the Commission by historian Paul Wallace Gates, and its final, public version runs over 800 pages. Anyone with professional interest in the history of the public lands will encounter this work and benefit from careful study.

The second is the Commission's final report, published in 1970. Titled *One Third of the Nation's Land*, it is a sweeping volume that covers all the major policy issues regarding public land use. Because it was written for the purpose of aiding congressional deliberations, it includes a long, detailed set of recommendations, some of which were adopted by Congress, some of which simply gathered dust, and many of which are by now quite dated.

These works are now well over half a century old, yet they provide an extraordinary snapshot of the state of public land law just as that law was about to be remade and modernized. They also reflect a sincerity and faith in government that today would appear naïve. The absence of cynicism or partisanship is striking. These documents remain fascinating and worthwhile reading, and much of the content is just as applicable today as at the time of its publication.

Congress responded to the environmental movement with a bipartisanship that is hardly imaginable today. The resulting environmental legislation of the 1960s and 70s was truly pathbreaking, and it had enormous implications for public land management. But it's worth noting that, for the most part, the enactments of this era did not repeal the laws that existed at the time. Instead, Congress simply added new layers on top of existing public land law. These layers accomplished far more than simply adding more national parks or monuments. They signaled an ecological turn in public land management, imposing many new requirements common to all four public land systems.

An Essential Clarification: Legislation vs. Regulation

To this point in this book, we have focused primarily on Congress and the laws it passed to govern public lands. After all, the Constitution vests Congress with the power to legislate, and congressional legislation is the starting point for almost everything that federal land agencies do. (Indeed, the agencies preface countless documents, including their annual budget requests, with a recitation of the statutes that authorize their activities.)

So Congress makes the law—but Congress can delegate regulatory authority to federal agencies. Those agencies then write rules that have the force of law. This is the distinction between legislation and regulation. Legislation is written by Congress; regulations are written by agencies (when they're authorized to do so) in a process called rulemaking. Rulemaking is a time-consuming process, but when that process is complete, the published rules are binding upon agencies and the general public.

When Congress delegates power to agencies, it often does so in very broad, general terms. This can give agencies a great deal of power because they decide how a broadly-written law will actually work. By extension, this also gives the president a great deal of power because the president ultimately oversees most federal agencies.[1] Newly-elected presidents often use their power to demand that agencies rewrite rules they oppose. Of course, Congress can step in at any time and pass a law that would override (or lock in) any particular rule. But the president can veto such legislation, and, in any event, Congress tends to move slowly. In effect, a new president can cause significant changes in the content and implementation of the law, even though the text of the relevant legislation hasn't changed at all.

For several decades now, Congress has been even more gridlocked than usual. Republicans and Democrats find it difficult to agree about anything. This makes the president's power over federal agencies even more important. If Congress is at an impasse and refuses to enact a president's agenda, the only way to advance that agenda may be through rewriting regulation. Pressured by the president to change a particular policy, federal agencies sometimes push their powers to the extreme, testing the limits of the legislation that authorizes them. This often leads to time-consuming lawsuits. It's very common for agencies to be sued, and one recurring complaint is that they have exceeded or abused their authority under the law that Congress has written.

BOX 4.2 Organic Legislation

Agencies don't emerge out of thin air; they must be created by Congress. The legislation creating an agency usually provides its marching orders, which are often updated by additional legislation in later years. The statute dictating the functions of an agency is often referred to as its "enabling act" or its "organic legislation."[2] Here is the organic legislation for each of the major agencies covered in this book:

- For the Forest Service, the *National Forest Management Act.*
- For the Bureau of Land Management, the *Federal Land Policy and Management Act.*
- For the National Park Service, the *National Park Service Organic Act.*
- For the Fish and Wildlife Service, the *National Wildlife System Improvement Act.*[3]

These organic acts provide a framework for each agency, but they do not direct every aspect of its work. Congress has also passed thousands of other statutes addressing the public lands. Some are quite specific and apply only to a single agency. For example, there is a statute that tells the Bureau of Land Management how to manage reindeer grazing on the Seward Peninsula in Alaska. We'll address some of these laws in the agency-specific chapters that follow. Other statutes address management by all four federal land agencies. As noted above, these statutes are our focus in the rest of this chapter.

All of this makes it important that as you read this book (or, indeed, any books or articles about the public lands) you pay careful attention to the distinction between acts of Congress, agency rules, presidential actions, and judicial decisions. Each of these kinds of federal action carries its own benefits and liabilities.

For the moment, our focus is on legislation—acts of Congress—and specifically on the raft of legislation Congress passed as the environmental movement took shape. As noted above, three laws deserve the most attention: the Wilderness Act of 1964, the National Environmental Policy Act of 1969, and the Endangered Species Act of 1973. The Wilderness Act is relatively straightforward, but the other two acts and their implementation have become exceedingly complicated. Let's dive in.

The Wilderness Act of 1964

Many public lands are not exactly pristine. Mines, ski resorts, timber harvests, and even national park visitor centers can significantly impair the ecological value of the lands on which they are located. In 1964, Congress passed legislation to require an extraordinary level of protection for public lands formally designated as protected wilderness. Under the Wilderness Act of 1964, wilderness is the most restrictive land use classification in all of public lands law. The Act's definition of wilderness has become iconic and can often be found emblazoned on signs throughout the federal land system:

> A wilderness, in contrast with those areas where man and his works dominate the landscape, is hereby recognized as an area where the earth and its community of life are untrammeled by man, where man himself is a visitor who does not remain.[4]

The definition goes on to state that wilderness should appear "to have been affected primarily by the forces of nature, with the imprint of man's work substantially unnoticeable" and that wilderness should provide "outstanding opportunities for solitude or a primitive and unconfined type of recreation."

What, specifically, is prohibited in wilderness areas? Almost everything. Inside a wilderness area, the Act bars commercial enterprise, permanent roads, motor vehicles, and even motorized equipment (although these last two are allowed in limited circumstances). Wilderness means wilderness. No mining, no logging, no buildings—not even chainsaws are allowed.[5]

Because wilderness is so restrictive, wilderness designations are made with great care. They can only be made by Congress, which means that they are often a part of a negotiated political compromise. For example, legislation creating a new wilderness area may also contain a policy concession to allay the concerns of opponents to the designation. The Wilderness Act itself invites Congress to allow pre-existing land uses to continue, even though they wouldn't ordinarily be permitted under the Act. Similarly, wilderness designations typically include language protecting "valid existing rights" in order to make clear that Congress is not extinguishing such rights.

Compare Figure 4.1 with Figure 2.1, and you'll note that only a small fraction of federal land has formal, legal wilderness status—about 5% (and about 2.7% of public land in the Lower 48). But the Wilderness Act packs a bigger punch than those statistics might suggest. This is because public land agencies manage some non-wilderness as though it might one day become wilderness. If an area is being considered for wilderness protection, the managing agency generally seeks to preserve that area's wilderness attributes. The Bureau of Land Management's maps, for example, sometimes display "Wilderness Study Areas." These are areas the Interior Secretary has recommended to Congress for wilderness protection but which have not (yet) received that formal designation. While the Secretary awaits Congress's decision, these areas are usually managed as though they were wilderness.

The National Environmental Policy Act (NEPA)

The National Environmental Policy Act of 1970 (NEPA; pronounced "NEE-puh") is one of the most important federal laws for public lands managers. It is just a few pages long, but its significance is difficult to overstate. And yet it is not what its name suggests: it is not a

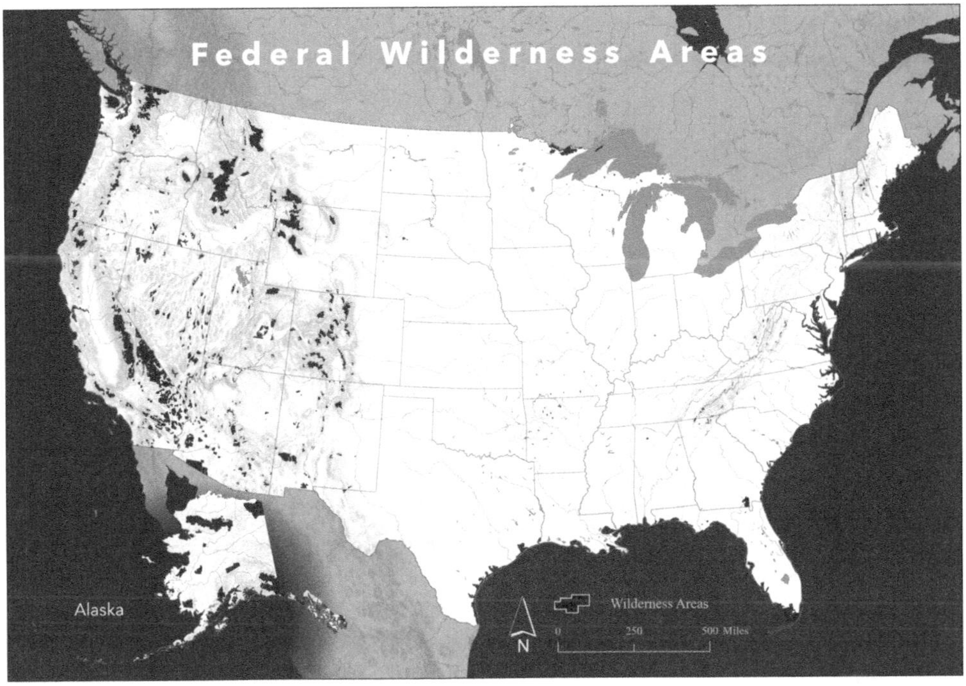

Figure 4.1 Federal wilderness areas. Map by Addison Arendt.

law that binds the government to a national environmental policy. To be sure, the opening paragraphs of the Act seem to set forth an ambitious policy. Its first sentences claim that its purpose is to "declare a national policy which will encourage productive and enjoyable harmony between man and his environment."[6] A moment later, the law announces some lofty goals for federal action: to "fulfill the responsibilities of each generation as trustee of the environment for succeeding generations" and to "assure for all Americans safe, healthful, productive, and aesthetically and culturally pleasing surroundings."[7]

But for better or worse, the courts have decided these grandiose phrases don't actually require anything in particular. Congress may have been sincere, but in the eyes of the federal judiciary, its language was not specific enough to impose any binding obligations on agency officials. No court of law has ever ordered the government to do (or not do) anything on account of the inspiring exhortations in NEPA's first paragraphs.[8]

Still, NEPA has muscle. It is hidden just a bit later in the statute. Section 102 of the law orders all federal agencies to:

> … include in every recommendation or report on proposals for legislation and other major Federal actions significantly affecting the quality of the human environment, a detailed statement by the responsible official on –
>
> (i) the environmental impact of the proposed action,
> (ii) any adverse environmental effects which cannot be avoided should the proposal be implemented,
> (iii) alternatives to the proposed action,

Figure 4.2 President Richard Nixon signing the National Environmental Policy Act into law on New Year's Day, 1970. Source: National Archives.

(iv) the relationship between local short-term uses of man's environment and the maintenance and enhancement of long-term productivity, and
(v) any irreversible and irretrievable commitments of resources which would be involved in the proposed action should it be implemented.

Unlike the general policy statements at the beginning of NEPA, Section 102 contains a clear, binding instruction that applies to "every recommendation" concerning a "major Federal action" which "significantly affect[s] the quality of the human environment." Certainly, some of these words are vague and require additional detail (such as "major" and "significant"), but they have real bite; federal courts routinely rebuke federal agencies for failing to produce the "detailed statement" demanded by Section 102 of NEPA. The purpose of the Section is to require agencies to look before they leap—to consider environmental impacts before they act.

The "detailed statement" has come to be known as the Environmental Impact Statement (EIS), and every federal land management agency produces a number of EISs every year. In a moment, we'll examine how EISs work and why they are so important. But first, we must make one more thing very clear. Under the current interpretation of the law, Section 102 of NEPA does not actually require any agency to act in an environmentally-friendly way. All it requires agencies to do is prepare a report that meets the requirements listed above. A report—that's it; that's all. Suppose an agency is deciding between two ways of building a road, one of which is environmentally destructive while the other is environmentally benign. Nothing in NEPA requires the agency to choose the latter. All the agency must do is write an adequate report about the environmental impacts of the two possible road plans; having done that, it's free to choose the environmentally destructive course if it wishes.

At this point, you may be wondering why NEPA is so important if its lofty policies aren't actually binding and if it doesn't compel agencies to limit environmental destruction. This is a great question. NEPA is important for several reasons. First, it requires agencies to do their homework and to share their analyses with the public by publishing EISs. Merely conducting an analysis may guide the agency toward a more prudent decision, and the transparency of the process gives the public an opportunity to provide input.

The second reason is more pungent. If EISs don't meet NEPA's requirements, agencies can be sued and forced by a court to re-do their work. And they are sued quite frequently. Environmental organizations monitor agencies' NEPA reports very, very closely. If an agency is considering an action that those organizations consider imprudent, you can bet that a NEPA-based lawsuit will follow. In fact, NEPA lawsuits are the most common kind of lawsuit faced by federal land managers—so much so that NEPA has become something of a punching bag for politicians. As we'll see momentarily, there are many who would like to dismantle NEPA's requirements.

But why would an environmental group sue an agency under NEPA, if all the law requires is a report? After all, if the agency loses, all it has to do is go back and write a better report, right? This is true, but EISs can take a long time to complete, and lawsuits can take years to resolve. A lawsuit based on NEPA may delay a federal action for a long time—and that delay is often all it takes to cause the agency to reconsider the action. Perhaps a key sponsor or participant in the project will abandon their support. Or perhaps a presidential election will occur in the meantime; if the new president opposes the contested action, he or she can now reverse course thanks to the delay caused by the lawsuit. These delays and their related expenses are the main reason why NEPA has such fierce opponents.

NEPA applies to all federal agencies, but obviously some agencies make more decisions affecting the environment than others. The Internal Revenue Service (IRS), for example, is almost never required to prepare an EIS, whereas the Army Corps of Engineers writes many EISs for river and flood control projects. And even in federal public land management, some agencies issue more EISs than others. The Forest Service and the Bureau of Land Management are among the busiest agencies when it comes to environmental assessment, precisely because they authorize a wider range of activities on the lands they manage: timber sales, mineral leases, leases for ski resorts, and so on. The Fish and Wildlife Service and National Park Service also prepare EISs, but far fewer.

BOX 4.3 The Council on Environmental Quality

To implement its objectives, NEPA created an entity within the Executive Office of the President called the Council on Environmental Quality (CEQ). In 1978, 8 years after NEPA was signed into law, CEQ issued regulations about how to implement the all-important "detailed statement" requirement of Section 102. The main purpose of these regulations was to flesh out the EIS process: to help agencies decide whether an EIS was necessary, what an EIS should include, and how an EIS should be incorporated into an agency's final decisions.

The 1978 rules remained in place and basically unchanged until 2020. In that year, the Trump Administration's CEQ issued a major overhaul to the NEPA rules, arguing that the NEPA process had become far too time-consuming and expensive. NEPA itself hadn't changed, of course, but the Trump Administration argued that its more relaxed rules were still perfectly consistent with the law. Unsurprisingly, when President Biden took office in January 2021, his administration vowed to undo most of the changes made in 2020. In May 2022, new CEQ rules took effect, returning EIS practice almost entirely to the 1978 rules. Then in February 2025, the second Trump Administration purported to repeal the CEQ regulations altogether.

Here's an example of how just one provision changed over this timeframe:

- **The 1978 rules** instructed agencies to consider not only *direct* environmental effects of their actions but also *indirect* and *cumulative* effects—effects which "when added to other past, present and reasonably foreseeable future actions" could be "individually minor but collectively significant ... over a period of time."
- **The 2020 rules** made analysis of *indirect* effects optional and deleted reference to *cumulative* effects.
- **The 2022 rules** restored the 1978 requirements as to indirect and cumulative effects. The only trace remaining of the 2020 changes is that agencies are required to consider only those effects that are "reasonably foreseeable," a limitation that didn't exist in the 1978 rules.
- **The 2025 repeal** eliminated the CEQ rules in their entirety, leaving each agency to implement NEPA for itself and to decide which environmental effects required study.

Until American politics undergoes its next major transformation, it's entirely likely that each successive president will change, resurrect, or repeal the NEPA rules yet again. The CEQ and its regulations have become political footballs.

The Environmental Assessment Process

Let's now look more closely at how the formal NEPA process plays out within agencies. Suppose that one of the chief Regional Foresters in the U.S. Forest Service is considering a timber sale to a forest products company. But suppose the Forester has some wiggle room under the law: she can sell as much as 400 million board feet (mbf), or she can sell less, or none at all.

Because of NEPA, the Forester can't simply make a decision and be done with it. Only simple, low-level decisions could be made without analysis. If the Forester was buying paper clips or hiring a maintenance worker, it would be quite clear that NEPA would not apply. For activities that are slightly more environmentally questionable, each agency publishes a lengthy list of activities that it performs regularly but that don't rise to the level of a "major" action with "significant" environmental effects. These activities are called "Categorical Exclusions" (CEs or "Cat-X's") because they are categorically excluded or exempted from further NEPA analysis. The publication of the list of CEs constitutes an act of regulation and must comply with the rules about rulemaking discussed above. You can find each agency's CEs published in the Code of Federal Regulation (CFR). Here is a sampling of some of the Forest Service's CEs:

> **36 CFR § 220.6 – Categorical Exclusions.**
>
> (d)(3) Repair and maintenance of administrative sites. Examples include but are not limited to:
>
> (i) Mowing lawns at a district office;
> (ii) Replacing a roof or storage shed;
> (iii) Painting a building; and
> (iv) Applying registered pesticides for rodent or vegetation control.

It's not surprising that mowing lawns or painting doesn't require a lot of analysis. But for any decision that even might significantly affect environmental quality, the Forester will have to undertake an environmental assessment (EA). The purpose of the EA process is to determine whether a full-blown EIS is required. Because NEPA only requires a complete EIS when there is a "major federal action" that "significantly affects the quality of the human environment," agencies must have a process to determine whether those thresholds have been met. In the majority of cases, the EA is the end of the road; the EA will determine that there is no significant impact, and the agency will simply publish a "Finding of No Significant Impact" (FONSI) and be done with it. In other instances, the agency will find that it can avoid the EIS threshold by taking some mitigating action. For example, suppose that a timber sale of 200 mbf would not trigger an EIS so long as the relevant timber is cut selectively rather than clear-cut. Our hypothetical Forester could approve the timber sale, contingent on taking a particular mitigating step; namely, requiring a selective cut. In this case, we'd call the agency's publication a "Mitigated FONSI." The finding of "no significant impact" is contingent on the agency's commitment to mitigation.

Collectively, federal agencies issue roughly 50,000 CEs each year. They perform perhaps another 3,000 EAs each year.[9] Of those EAs, only a small percentage will lead to a complete EIS, and the total number of EISs has dropped in recent years. We turn now to the EIS process.

The Environmental Impact Statement

A complete EIS is a behemoth of a document. In fact, it's almost never just one document; an EIS is typically a collection of multiple documents, which together might run to thousands of pages, complete with appendices, charts, and other supporting evidence. The final collection of documents might not even be labeled as an EIS—the typical designation for that collection is often the "Record of Decision" (ROD).

The heart of the ROD or EIS is typically its "alternatives analysis," the analysis of the different options the agency considered. It wouldn't be practical to analyze every possible alternative, so agencies typically review a handful of choices that are varied enough to allow serious consideration of the major differences. One of the alternatives is the "no action alternative"—that is, not doing anything at all (unless that's simply not an option under the governing law). After spelling out the various alternatives in sufficient detail, the agency proceeds to analyze the environmental impacts that would result from each option.

Back to our hypothetical Forester. Suppose the Forester leaned toward the largest timber sale possible in her situation, a sale of 400 mbf. Suppose also that the EA process determines that an EIS is necessary. The ROD might assess three options—a sale of 400 mbf, a sale of 200 mbf, and no sale at all (the "no action alternative")—in terms of their environmental impacts.

If the environmental analysis is not thorough and professional, there is a good chance the agency will be sued. (In high-profile instances, the agency would likely be sued even if the EIS was as thorough and rigorous as possible.) The ever-present risk of litigation has caused a cottage industry of environmental consultants to emerge; today, a significant proportion of EIS work is not handled by the agency itself but is farmed out to these consultants.[10] Still, there is no guarantee that an EIS will survive a legal challenge simply because it was written by professional consultants!

What are the sorts of NEPA violations that cause federal courts to reject EISs? It could be that the agency failed to consider certain scientific publications or a particular kind of environmental impact. For example, many agencies long refused to analyze how their actions would affect climate change. This was not necessarily borne out of political opposition or climate denialism; instead, the agency might have reasoned that its action was miniscule as compared to the world's cumulative emissions. But in recent years, some courts have required agencies to include an analysis of climate impacts if the contested agency action involves, say, the production of fossil fuels on federal land. And there are numerous other ways that an EIS can find legal fault in the eyes of a judge.

Ongoing Disputes About NEPA

Although many environmentalists have grown attached to NEPA, there is an ongoing debate about whether NEPA's requirements, as presently implemented, go too far. The NEPA process has become so onerous and so lengthy that even many environmentalists have become concerned. Many renewable energy projects, for instance, require transmission lines or pipelines across federal lands. These projects often require a complete EIS, which can add years to the project's timeline. According to the Department of Energy, between 2010 and 2018 the average EIS took 4.5 years to complete. The Council on Environmental Quality estimates that the typical EIS costs between $250,000 and $2 million to prepare.

Today, numerous constituencies favor some sort of NEPA reform. But it's unlikely that Congress will repeal NEPA altogether. Almost certainly, public land managers for decades to come will still have to make difficult decisions about environmental analysis.

The Endangered Species Act (ESA)

The Endangered Species Act of 1973 is the other major federal statute that every federal land management agency confronts on an almost daily basis. It is a potent law; some have called it "the pit bull of environmental law." The reason for this potency has to do with the law's absolute terms. In stark contrast to NEPA, which contains much ambitious but ultimately toothless language, the ESA has sharp teeth indeed—so sharp, in fact, that federal agencies are sometimes reluctant to bite as powerfully as they might for fear of provoking opposition to the law.

Like NEPA, the ESA was signed into law by President Nixon. There is little indication that Nixon or members of Congress realized how powerful the law might be.[11] That realization settled in just a few years later, in one of the most important environmental law cases the Supreme Court has ever decided.

Tennessee Valley Authority v. Hill (1978)

In the 1960s and 70s, the federal government was still in dam-building mode. New dams today have fallen out of favor for a variety of reasons, but at that time, dams were regarded not only as a source of clean, cheap hydropower but also as a kind of economic stimulus. Building dams created jobs, and just as importantly, dams created reservoirs and thus many miles of new lakefront property. In this instance, the TVA wanted to build a dam on the Little Tennessee River. Although the dam would inundate productive farmland (along with homes and other buildings), the TVA believed that the inexpensive power and higher-value real estate that would result would be an economic boon to the region.

By 1975, the Tellico Dam, as it was called, was over halfway done. There was just one problem: a new species of fish, the tiny snail darter, had been discovered in these very same reaches of the Little Tennessee—and nowhere else. Some enterprising plaintiffs, and the scientists who supported them, argued that the snail darter could well become extinct if the dam was completed. They sued the TVA, relying on the ESA's requirement that agencies shall:

> … insure that actions authorized, funded, or carried out by them do not jeopardize the continued existence of [ ] endangered species and threatened species or result in the destruction or modification of habitat of [ ] species which is determined by the Secretary … to be critical.
>
> (Endangered Species Act, Section 7)[12]

Read those words carefully. Taken literally and at face value, they disallow agencies from so much as modifying any habitat that is "critical" for an endangered or threatened species. The local District Court (the first court to hear the lawsuit) recognized that the Tellico Dam would "result in the adverse modification, if not complete destruction, of the snail darter's critical habitat," making it "highly

Figure 4.3 The Tellico Dam during construction in 1967. Source: Tennessee Valley Authority.

probable" that "the continued existence of the snail darter" would be "jeopardize[d]."[13] But the court simply couldn't stomach the idea that a tiny fish could stop the federal government from finishing a multi-million dollar dam project that was nearly complete. Here is how the U.S. Supreme Court described the District Court's reasoning:

> [T]he District Court stressed that the entire project was then about 80% complete and, based on available evidence, "there [were] no alternatives to impoundment of the reservoir, short of scrapping the entire project." The District Court also found that if the Tellico Project was permanently enjoined, "some \$53 million would be lost in nonrecoverable obligations," meaning that a large portion of the \$78 million already expended would be wasted …
>
> To accept the plaintiffs' position, the District Court argued, would inexorably lead to what it characterized as the absurd result of requiring "a court to halt impoundment of water behind a fully completed dam if an endangered species were discovered in the river on the day before such impoundment was scheduled to take place. We cannot conceive that Congress intended such a result."[14]

But the Supreme Court ultimately disagreed with the District Court. The pivotal language is worth quoting in full:

> It may seem curious to some that the survival of a relatively small number of three-inch fish among all the countless millions of species extant would require the permanent halting of a virtually completed dam for which Congress has expended more than $100 million. The paradox is not minimized by the fact that Congress continued to appropriate large sums of public money for the project, even after congressional Appropriations Committees were apprised of its apparent impact upon the survival of the snail darter. We conclude, however, that the explicit provisions of the Endangered Species Act require precisely that result.
>
> One would be hard pressed to find a statutory provision whose terms were any plainer than those in § 7 of the Endangered Species Act. Its very words affirmatively command all federal agencies "to *insure* that actions *authorized*, funded, or *carried out* by them do not jeopardize the continued existence" of an endangered species or "*result* in the destruction or modification of habitat of such species … ." 16 U.S.C. § 1536 (1976 ed.). (Emphasis added.) This language admits of no exception.[15]

"This language admits of no exception." With that sentence, the ESA became the pit bull it is today. The Supreme Court explained clearly that it would apply the law exactly as it was written. The Act implicates every federal agency and every action they authorize, fund, or carry out. The responsibilities imposed by the Act weigh especially heavily on the agencies that manage our public lands.

The Endangered and Threatened Species Lists

So let's examine the ESA and its core provisions. You've just read a short but powerful excerpt from Section 7 of the Act, which made reference to endangered and threatened species. That language refers back to Section 4, which instructs the U.S. Fish and Wildlife Service (FWS)—the same agency that manages National Wildlife Refuges—to maintain both an Endangered Species List and a Threatened Species List. An endangered species is one that is "in danger of extinction throughout all or a significant portion of its range," whereas a threatened species is one that is likely to become endangered "in the foreseeable future."[16] For our introductory purposes, the differences between these lists are not important; both categories receive a great deal of legal protection under the ESA.

Maintaining these lists is one of the most important and contentious tasks performed by the FWS. It requires a great deal of scientific and administrative effort. The FWS can nominate a species for protection on its own initiative, but the law also allows others to petition the agency for listing. Once a petition has been received, the agency has just a year to determine whether to list it. The law requires the FWS to make this determination on the basis of the "best scientific and commercial information available." After reviewing this information and researching the matter thoroughly—which is a major operation!—the agency may decide to add the species to one list or the other, to refuse to list the species, or to refrain from listing because of inadequate information or because listing is "warranted but precluded" by higher-priority species.

At the time of this writing, there are about 1,300 endangered and threatened species. A sizeable proportion of these species exist on the public lands, and as we'll see, that fact creates important duties for the managers of those lands.

Protections for Listed Species

Adding a species to the Endangered or Threatened Species Lists triggers a host of important requirements which affect both the federal government and the general public. Members of the public, for example, face criminal penalties for "taking" a listed species of fish or wildlife—and taking is defined very broadly by the Act to include not just killing or capturing but also harassing, harming, pursuing, or hunting a member of the species. For its part, the FWS's reward for the hard work of listing a species is … more hard work. When a species is listed, the FWS is required to figure out the boundaries of that species' "critical habitat" and to prepare a recovery plan to bring the species away from the brink of extinction.

Our focus here, though, is on the ESA rules that directly bind public lands managers. As you've seen, Section 7 of the ESA requires them to "insure that actions authorized, funded, or carried out by them do not jeopardize the continued existence" of listed species or "result in the destruction or modification" of critical habitat. Since many listed species find their habitat on public lands, these obligations can be quite burdensome. Simply figuring out whether a proposed action will jeopardize a species or modify its habitat can take years of scientific research. Typically, the ESA's research and administrative tasks are simply added to the NEPA review process, and together these tasks add an enormous procedural layer to federal land management. As we saw with NEPA, failure to properly dot i's or cross t's on these analyses can create a litigation nightmare. Even extensive, careful analyses may trigger lawsuits designed to delay or frustrate projects.

Examples of Endangered Species Issues on Public Lands

The best way to get a sense of the importance of endangered species to public lands management is through actual events and situations, like the story of Tellico Dam and the snail darter. Another similar story involves the northern spotted owl and timber harvesting in the Pacific Northwest, an area renowned for its incredibly valuable old-growth forests. The forest products industry has long been vital to Washington and Oregon, thanks to the vast forests on the western side of the Cascade Range, many of which include abundant acreage in public ownership. Timber harvests from large-diameter, old-growth trees, such as those in these forests, are the most profitable in the country.

So when it came to light that forests in the Pacific Northwest constituted the habitat of the northern spotted owl, whose population was declining precariously, it sent a shock wave through the entire regional economy. Forest Service officials recognized immediately that continued timber sales could be imperiled on account of Section 7's rules, and they said so publicly. Lumber workers responded with fury, as did many of those who worked in sectors downstream from timber production. Outrage over the ESA quickly took over the local news cycle. Communities feared the loss of vital local employers, such as sawmills and paper factories. Bumper stickers popped up encouraging the hunting of spotted owls out of spite.

The dispute became so heated that ultimately the Clinton Administration arranged a summit involving the governors of Washington and Oregon, the Chief of the Forest Service, and numerous other VIPs to broker a deal. To make a very long story short, the spotted owl (and the ESA itself) survived; timber harvests from the Northwest's public lands were dramatically reduced (see Figure 4.4); and the economies of these two states shifted away from forest products and toward high-tech and other industries.

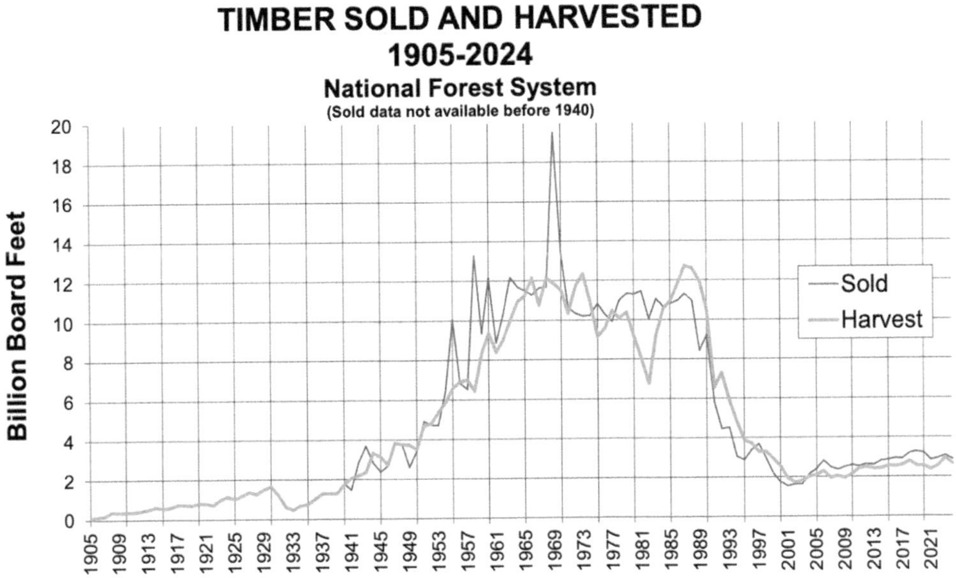

Figure 4.4 Timber sales and harvests from USFS lands, 1905-2024. Note the precipitous decline in the early 1990s, caused by decreased production in the Pacific Northwest. Source: Forest Service.

As we'll see later in the book, however, resentment from this episode lingers in ways that continue to affect public land management some three decades later.

Another species that has bedeviled public land managers is the grey wolf. Wolves are simultaneously loved and hated—on one hand, an icon of rugged and fierce survivalism; on the other, a muscular nemesis and a savage predator of domesticated wildlife. A public lands manager might encounter, on the same day, a demand to manage land to prevent wolves from attacking livestock on adjacent ranches, as well as a lawsuit from an environmental group arguing that certain trails must be closed due to human disturbance of wolf mating behaviors.

Although most Americans favor protections for wolves, significant disagreements persist. Whether wolves are actually endangered is itself a matter of dispute. First listed in 1974, they were delisted by the Trump Administration in 2020—a move that the Biden Administration also defended. Yet a federal court reversed the delisting, maintaining wolves' endangered status. Hunters tend to oppose wolf recovery, in part because wolves predate on species that hunters prize, such as moose. Even more broadly, attitudes about wolves reflect a growing urban–rural divide about the very meaning of conservation. Urbanites tend to prefer policies that protect nature for its own sake and that avoid harming any individual species, such as wolves. Rural residents, by contrast, are more likely to prioritize human well-being, and thus may endorse the hunting of wolves in order to protect humans and domesticated animals.

One final example involves the greater sage grouse (see Figure 4.5). This bird's habitat extends across a vast portion of the Mountain West. Its populations are in decline, but whether the species is truly endangered is very much in dispute. More to the point, its listing could significantly affect all manner of activities in this region—from mining and oil and gas production to recreation and homebuilding. For this reason, Congress itself

Figure 4.5 The greater sage grouse. Source: Tom Koerner, Fish and Wildlife Service.

stepped in after more than a decade of conflict over the grouse's status. For several consecutive years, Congress passed appropriations riders forbidding the FWS from spending any funds to conduct any ESA-related analysis of the grouse. The agency was forced to terminate the listing process, but the episode also pressed western states into action. State governments realized that it was in their own interest to protect the grouse through state policy—because unless grouse populations improve, an ESA listing looms.

This is a prime example of how, thanks to the stringency of the ESA, particular species have a tactical relevance quite apart from their conservation status. Some wildlife advocates sought a grouse listing precisely because of the bird's extensive range and because of the anti-development effect a listing would bring. Proponents of development were willing to support state-level conservation programs they'd otherwise oppose to prevent a listing. Such is life under the Endangered Species Act!

Putting It All Together: The Federal Land Planning Process

The laws described above are complicated, and we've only scratched the surface. There are many more laws we haven't discussed—including some vastly important environmental statutes, such as the Clean Air Act and the Clean Water Act—that significantly affect portions of the federal estate. Managing the public lands is a legally-intensive task. And each of these laws exists because Congress passed it, which is to say that federal lawmakers perceive that the public wants these laws.

So how do all these laws fit together? How can a public land official be sure that he or she is complying with the law? The answer is process—a careful, meticulous process that requires officials to think carefully about the particular land they manage and the laws that bear on those lands. The principal federal agencies discussed in this book—the National Park Service, the U.S. Forest Service, the Fish and Wildlife Service, and the Bureau of Land Management—all undertake a planning process and produce management plans, which are usually updated every several decades. These plans incorporate the legal obligations described above and many others besides. The particular requirements of the planning process are often spelled out in agency-specific rules, some of which we'll encounter in the chapters that follow.

Management plans can be tricky documents. They identify an agency's goals at a moment in time, but those goals can change overnight. Plans can provoke or quell lawsuits. They can lash agencies to the proverbial mast, making it more difficult for them to radically change course. And plans can also be boring and tedious, laden with the minutiae of land management. We could easily fill an entire volume with a discussion of each agency's planning rules, but we can't spend more time on the details of the land planning process here without turning this into quite a different book. Whatever else we might say about it, one thing is certain: land management planning takes time. It represents an extraordinary burden on our public land managers. But it is a burden that We the People have imposed with the weighty instrument of the Law of the Land.

BOX 4.4: Find A Plan, Any Plan

Before you read further, you may wish to take a look at an actual management plan. They are quite easy to find because federal agencies post them on their websites. Search the web using the name of any unit of public land, along with the words "management plan" or their equivalent. With just a couple of clicks, you will be looking at a plan, whether it's the "1986 Theodore Roosevelt National Park General Management Plan" or the "Craters of the Moon Resource Management Plan" or the "2019 Inyo National Forest Plan." You'll gain an immediate sense of the kinds of legal issues that confront the managers of that particular land unit. In all likelihood, you'll find repeated references to NEPA, the ESA, and perhaps the Wilderness Act. And whatever issues are discussed, you'll see right away that these plans are extensive and detailed, prepared with great effort and diligence by hard-working public servants.

Conclusion

Since the early days of the United States, federal lawmakers have been preoccupied with the public lands. Congress wrote laws to acquire land, to dispose of land, to retain land, and to make use of land. The modern environmental movement demanded its own body of law, requiring public land managers to preserve wilderness, to study the environmental impacts of various land uses, and to protect endangered and threatened species. Today's public land managers must be well-versed in the laws that guide their agencies and must engage in regular planning efforts to ensure compliance with those laws. In the coming chapters, we'll focus on each of the four major land management agencies in turn, beginning with the National Park Service.

Study Questions

1. In what ways did the environmental legislation of the 1960s and 70s introduce an "ecological turn" to public land management?
2. How does NEPA serve as a potent tool for environmental groups, despite not compelling environmental outcomes?
3. How does the ESA influence state-level conservation policies?

Resources for Further Study

Kalen, Sam. "NEPA's Trajectory: Our Waning Environmental Charter from Nixon to Trump?" *Environmental Law Reporter* 50 (2020): 10398–10422.

Nash, Roderick. *Wilderness and the American Mind* (New Haven, CT: Yale University Press, 1967).

Plater, Zygmunt. *The Snail Darter and the Dam: How Pork-Barrel Politics Endangered a Little Fish and Killed a River* (New Haven, CT: Yale University Press, 2013).

Public Land Law Review Commission. *One Third of the Nation's Land: A Report to the President and the Congress by the Public Land Law Review Commission* (U.S. Government Printing Office, 1970).

Sax, Joseph L. "The (Unhappy) Truth About NEPA." *Oklahoma Law Review* 26 (1973): 239–248.

Notes

1. Some agencies, such as the Federal Reserve or the Federal Trade Commission, have long been considered to be "independent," almost entirely free from presidential control. As this book goes to press, the Trump Administration is testing the limits of this independence.
2. This has nothing to do with *organic* in the sense of *environmentally friendly*, and everything to do with *organic* in the sense of *organize*.
3. This legislation is a bit different than the organic legislation for the other three agencies in that it speaks only to the National Wildlife Refuge System. The U.S. Fish and Wildlife Service has other important functions apart from its responsibility for the Refuge System, but these functions are not covered by the Improvement Act. Nonetheless, Rob Fischman has convincingly argued that the Act serves as the organic act for the Refuge System in all important respects. See Robert L. Fischman, *The National Wildlife Refuges: Coordinating a Conservation System Through Law* (Washington, D.C.: Island Press, 2003), 64–76.
4. 16 U.S.C. § 1131(c).
5. One of the research assistants for this book spent a summer doing trail repair in a wilderness area managed by the Forest Service. He was not allowed to use a power drill to fix a small wooden bridge.
6. 42 U.S.C. § 4321.
7. 42 U.S.C. § 4331.
8. In the United States, courts rarely find any enforceable content in broad statements of policy, especially when those statements provide some flexibility. For example, NEPA instructs the Federal Government to pursue its goals with "all practicable means, consistent with other essential considerations of national policy" (42 U.S.C. § 4331(b)). That clause alone provides more than enough wiggle room to avoid judicial sanction.
9. Statistics on EAs are hard to come by. This estimate comes from John C. Ruple and Kayla M. Race, "Measuring the NEPA Litigation Burden: A Review of 1,499 Federal Court Cases," *Environmental Law* 50 (2020): 479–522, 485–486.
10. For a very interesting study of these consultants, see Dave Owen, "Consultants, the Environment, and the Law," *Arizona Law Review* 61 (2019): 823–884.
11. Only four members of Congress—in the House and Senate combined—voted against the law.
12. 16 U.S.C. § 1536(a)(2).

13 419 F. Supp. 753, 757 (ED Tenn.), cited by the Supreme Court in *TVA v. Hill*, 437 U.S. 153, 165–166 (1978).
14 *TVA v. Hill*, 437 U.S. at 166–167.
15 *TVA v. Hill*, 437 U.S. at 172–173.
16 16 U.S.C. § 1532(6), (20).

5 The National Park Service

Introduction

This chapter will introduce you to the National Park Service (NPS) and some of the issues and conflicts that have shaped its approach to the management of public lands. As with each of the four primary land systems, we will begin with the lands and resources themselves. From there we turn to the mission of the NPS; to the policies, laws, and regulations that govern the agency; and finally to the agency's organization. By the chapter's end, you will have a basic understanding of the national park system that will help you differentiate it from the other three systems.

The Diversity of the National Park System

America's national parks are among the most treasured and beloved protected lands in the world. Marquee parks like the Grand Canyon attract millions of tourists each year, but the national park system also includes lands that few will ever see. This far-flung system reaches from the Arctic Circle to the Caribbean, touching nearly every biome and ecological zone in the United States. Protecting such diverse lands, while withstanding the pressures associated with mass tourism, is the NPS's task.

Pause for a moment and visualize Death Valley, Denali, Acadia, Yosemite, Yellowstone, the Great Smoky Mountains, or the Grand Canyon. (If you're unfamiliar, look at some pictures online.) These sites could not be more different, yet American law places them into the same public land system. The park system is best known for these "crown jewels," which are recognized around the world. Yet the system also includes far less familiar parks and hundreds of other units classified not as parks but as monuments, recreation areas, historic sites, and so forth.

Why such a hodgepodge? The national park system is a *political* creation. There is no panel of experts that hand-selects the most exotic or spectacular sites and declares them to be parks. Many of the country's most breathtaking landscapes are unprotected. For every Yosemite Valley, there is a Hetch Hetchy; for every Grand Canyon, there is a Glen Canyon. The Hetch Hetchy Valley is just a few miles and one watershed north of Yosemite Valley. Its granite theater is every bit as wondrous as its neighbor, but at its mouth now stands an enormous dam. Where there might otherwise be streams and footpaths winding through tree-ringed meadows, there is instead a reservoir drowning the valley in hundreds of feet of water. America's archetypal naturalist, John Muir, labored mightily but unsuccessfully to prevent the dam's construction. A year after he died, Congress authorized the dam as a means of providing a municipal water supply

DOI: 10.4324/9781003392514-5

to the city of San Francisco. Similarly, the sprawling and transcendent red-rock reaches of Glen Canyon, just one hundred miles up the Colorado River from the heart of the Grand Canyon, are now hidden under the waters of Lake Powell. The Glen Canyon Dam, constructed by the mighty Bureau of Reclamation during its midcentury heyday, was intended to forestall a water supply shortage for the growing desert cities of Arizona, California, and Nevada. The Hetch Hetchy and Glen Canyon dams provide important public benefits, but at the cost of inundating spectacular terrain. These types of decisions are quintessentially political, not technical.

Congress designates national parks. In this as in other areas of public policy, Congress is known more for its tendency to bargain, compromise, and cut deals than for its disciplined systematicity. And members of Congress are not solely concerned with the *national* interest; they have a duty—indeed, a political imperative—to represent their home states and to serve their constituents. When it comes to national park designation, state delegations have significant power. This helps explain why park units in one state might differ substantially from units in another. Few would contend that the Indiana Dunes National Park contains scenery on par with Rocky Mountain or Mount Rainier National Park, but when Indiana's representatives lobbied for the national park classification in 2019, legislators elsewhere had little reason to oppose it. In fact, their support for Indiana in this cause may have led Indiana legislators to support other states on other matters. As the saying goes, "If you scratch my back, I'll scratch yours!"

Over the last 150 years, Congress has established a curious collection of national park units to preserve everything from beautiful scenery to historic sites to cemeteries. Today most of the 433 park units fall into 19 distinct categories, listed in Table 5.1 below.[1]

Table 5.1 Units Administered by the National Park Service

National Battlefields	11
National Battlefield Parks	4
National Battlefield Sites	1
National Military Parks	9
National Historical Parks	63
National Historical Sites	76
International Historic Sites	1
National Lakeshores	3
National Memorials	31
National Monuments	87
National Parks	63
National Parkways	4
National Preserves	19
National Reserves	2
National Recreation Areas	18
National Rivers	4
National Wild and Scenic Rivers and Riverways	10
National Scenic Trails	6
National Seashores	10
Other Designations	11

Figure 5.1 Land managed by the National Park Service. Map by Addison Arendt.

Although we don't have space to address all of these categories, the paragraphs that follow will give you a sense of the diversity within the park system.

The Crown Jewels: Scenic and Ecological Parks

Perhaps a dozen parks stand out for their international fame, their iconic popularity with tourists, their vast scale, and their longstanding status as bucket-list destinations. This category includes parks such as Yellowstone, Yosemite, Grand Canyon, Glacier, Everglades, Great Smoky Mountains, Denali, Acadia, Olympic, Rocky Mountain, Zion, and Bryce Canyon. These are parks that Congress created primarily to protect natural scenery and wildlife habitat and to create unique opportunities for public recreation. Though most are remote, ample means exist by which millions of tourists may stream through these sites each year. Their ever-growing popularity creates difficult challenges for both the National Park Service and for nearby gateway communities. This challenge has only increased as the NPS has expanded its work beyond preserving scenery to a far more complex set of goals related to preserving ecological systems and processes. We will have much more to say about these sites and their associated challenges later in the book.

Cultural and Historical Sites

The crown jewel sites listed above are some of the earliest national parks, and they are largely renowned for their scenery. But there is another, more recent dimension to the NPS's work. The NPS is now one of the most important curators of the nation's cultural and social history. Over the past several decades, the agency has taken up the task of helping visitors view its lands not simply as scenic playgrounds or geologic curiosities but as sites that bear witness to profound and sometimes painful human experiences. Indigenous culture and the social history of the U.S. are now commonly narrated in interpretive materials throughout the park system.

Here is a sampling of the several hundred sites within the park system that memorialize aspects of America's cultural, social, and military history.

- The Manzanar National Historic Site, in a remote part of east-central California, memorializes the incarceration there of some 120,000 Japanese-Americans during World War II.
- The 42-acre River Raisin National Battlefield Park in southeast Michigan is the only park system property devoted to a site of the War of 1812.
- The *Brown v. Board of Education* National Historical Park in Topeka, Kansas, commemorates the 1954 decision of the U.S. Supreme Court to end racial segregation in public schools.
- In the heart of Washington, D.C., the NPS manages an array of memorials to Abraham Lincoln, Thomas Jefferson, the Vietnam War, the Korean War, World War I and II, etc.

Sites like these require the Park Service to serve as an official curator of America's social memory. At first glance, this may seem like an easier task than ecological protection. But interpretations of history are contested and change regularly, and the NPS must navigate this shifting interpretive landscape. For example, consider the agency's presentation of the

Little Bighorn Battlefield National Monument, where General Custer and 263 soldiers died fighting a combined force of Arapaho, Lakota, and Cheyenne warriors in 1876. It was originally declared a national cemetery to protect the soldiers' graves in 1879. The National Park Service assumed responsibility in 1940, and the site was redesignated Custer Battlefield National Monument in 1946. The NPS largely celebrated General Custer and his men as heroes, with little honor given to the Indians who died in the battle. But after the Civil Rights Movement and the American Indian Movement, the NPS shifted its interpretation of Little Bighorn. Congress adopted the current name in 1991, and the NPS dedicated an Indian Memorial in 2003. Today, the NPS highlights the injustices suffered by Native Americans as well as the bravery of the Indian warriors who fought and died in the battle.

Moreover, virtually *all* of the land now managed by the National Park Service was once occupied by Indigenous peoples. Between roughly the years 2000 and 2025, the NPS took large strides toward interpreting the landscapes it stewards with reference to Indigenous practices, history, and culture. This occurred even at sites most famous for their scenery. The task of educating visitors about America's Indigenous forebears—and the deep injustices they suffered at the hands of American government—extended across the entire National Park System.

With the reelection of President Trump in 2024, the pendulum began to swing back in the other direction. In an Executive Order, Trump directed the Interior Department, of which the NPS is part, to ensure that

> properties within the Department of the Interior's jurisdiction do not contain descriptions, depictions, or other content that inappropriately disparage Americans past or living (including persons living in colonial times), and instead focus on the greatness of the achievements and progress of the American people or, with respect to natural features, the beauty, abundance, and grandeur of the American landscape.[2]

Land managers in the NPS were once again caught in the middle of a broader social and political shift. You can be certain it will not be the last time!

National Monuments

National monuments are a fascinating and diverse category. Unlike national parks, which are created only by Congress, national monuments can be designated unilaterally by the president under the Antiquities Act of 1906. Most presidents since 1906 have used monument designations to burnish their record of conservation, especially in the closing days of their terms of office. Table 5.2 lists some of the most recently created national monuments within the National Park System and the presidents who designated them. (There are many other recent monuments besides those listed, but they are managed by other agencies.) What patterns do you notice?

Because presidents can designate monuments unilaterally, monuments represent a flexible "holding pattern" category into which lands can be placed indefinitely as they await possible reclassification. Over the years, many national monuments have become national parks—sometimes almost immediately, sometimes decades later. Other monuments have been transferred to states or to other federal agencies. But dozens remain in monument status in perpetuity. The majority of these are managed by the National Park Service, and many contain outstanding features comparable to those

Table 5.2 National Monuments Designated by Recent Presidents for NPS Management

President & Name of Monument	*Description*	*Date of Designation*
President Biden:		
• Frances Perkins	Residence of the first female cabinet member in U.S. history	Dec. 16, 2024
• Carlisle Federal Indian Boarding School	The leading Federal Indian Boarding School, open from 1879–1918	Dec. 9, 2024
• Springfield 1908 Race Riot	Site of mass racial violence	Aug. 16, 2024
• Emmett Till and Mamie Till-Mobley	Several sites related to infamous lynching, a pivotal moment in the Civil Rights Movement	Jul. 25, 2023
• Avi Kwa Ame	Mountain in the Mojave Desert, sacred to Indigenous tribes	Mar. 21, 2023
President Trump:		
• Medgar & Myrlie Evers Home	Home of civil rights activists	Dec. 10, 2020
• Mills Spring Battlefield	Civil War battlefield	Sept. 22, 2020
• Camp Nelson	Civil War recruitment center for African Americans	Oct. 26, 2018
President Obama:		
• Birmingham Civil Rights	Motel bombed by KKK in 1963	Jan. 12, 2017
• Freedom Riders	Sites where Freedom Riders were attacked in 1961	Jan. 12, 2017
• Katahdin Woods and Waters	Mountains and wilderness in Maine	Aug. 24, 2016
• Stonewall	Gay bar in NYC and site of riots in 1969	Jun. 24, 2016
• Belmont Paul-Women's Equality	Home of National Women's Party	Apr. 12, 2016
• Castle Mountains	Desert grasslands and mountains in Mojave Desert	Feb. 12, 2016
• Waco Mammoth	Mammoth fossils	Jul. 10, 2015
• Pullman	First "company town" in the U.S.	Feb. 19, 2015
• Tule Springs Fossil Beds	Prehistoric fossils	Dec. 19, 2014
• Charles Young Buffalo Soldiers	Home of first African-American colonel in the U.S. Army	Mar. 25, 2013
• César E. Chavéz	Home and gravesite of labor and civil rights leader	Oct. 8, 2012
• Fort Monroe	Civil War fort in Chesapeake Bay	Nov. 1, 2011

President & Name of Monument	*Description*	*Date of Designation*
President George W. Bush:		
• Tule Lake	Japanese internment center in World War II	Dec. 5, 2008
• African Burial Ground	Seventeenth- and eighteenth-century African-American burial ground	Feb. 27, 2006
President Clinton:		
• Virgin Islands Coral Reef	Coral reefs and mangrove forests	Jan. 17, 2001
• Grand Canyon-Parashant	Remote terrain near Grand Canyon with abundant fossils	Jan. 11, 2000
• Agate Fossil Beds	Prehistoric fossils	Jun. 14, 1997
President George H.W. Bush:		
• Petroglyph	Archaeological and volcanic sites	Jun. 27, 1990

within actual national parks. The other three land agencies manage a smaller number of monuments and manage them according to very different standards. The array of lands in this category mirrors the variety in the entire park system, from sublime landscapes to battlefields to civil rights memorials.

National Recreation Areas

Within the park system is a set of national recreation areas (NRAs), a classification employed by Congress since the 1960s. These areas tend to be large and open to an array of recreational uses (and thus all the impacts that accompany them). Many NRAs surround large bodies of water, such as Lake Mead, Lake Powell, Lake Chelan, and Lake Roosevelt. Others are adjacent to or even within urban areas; examples include the Golden Gate NRA (near San Francisco) and the Gateway NRA (near New York City). Due to their location and their recreational appeal, some of these units regularly rank among the most visited within the entire park system. In 2024, Golden Gate reported over 17 million visitors, and Gateway over 9 million.

Rivers, Lakeshores, and Seashores

Bodies of water (or portions of them) sometimes fall within the park system. Stretches of river designated as National Scenic Rivers, such as the Merced River in Yosemite National Park and the Snake River in Oregon and Idaho, gain additional legal protections beyond those conferred simply by park status. There are also several National Lakeshores, all of which are located on the Great Lakes, and ten National Seashores, mostly located along the Eastern Seaboard.

Older and Newer Parks

A useful distinction can be made between older parks and newer parks. Whereas the older park system units were carved out of the public domain (land that the federal government already owned), newer units had to be purchased or acquired from private landowners. Some of these land acquisition campaigns led to significant controversy. Congress has long authorized the Park Service to use eminent domain (the power of governments to force the sale of land for public use), but compelled sales can be deeply unpopular. In the 1970s, opposition to eminent domain actions threatened to erode public support for the creation of new park units, so the Service changed its tactics. In places like Cape Cod and Point Reyes, the Service negotiated a wide array of land transfer arrangements instead of condemning land outright. For example, the Service sometimes purchased only "future interests" in desired parcels, which allowed existing residents to remain for a negotiated amount of time. To this day, some park units contain properties on which private residents remain, even though title to the land will one day be transferred to the Park Service.[3]

Alaska

Alaska is in a class by itself. Its park system units are almost unfathomable in scale. Two out of every three acres within the park system are located in Alaska. The largest unit, the Wrangell-St. Elias National Park and Preserve, is 13.2 million acres in size, larger than Maryland, Massachusetts, and Delaware combined. The history of these units is also distinctive. Because Alaska is so vast and so sparsely populated, conflicts between settlers and conservationists, which often catalyzed parkland protection in the Lower 48, did not arise. Unresolved questions of land title in Alaska remained in a state of suspended animation until the discovery of massive oil reserves in the 1950s and 60s. With substantial oil revenue at stake, the state government, the federal government, and Native American tribes began jockeying for formal ownership of the state's 365 million acres—and for the money that would flow from leasing away the oil beneath.

With several momentous pieces of legislation in the 1970s and 80s, Congress carved up Alaska's unclaimed land. The state government received over 100 million acres; Native American tribes got 50 million acres, along with a considerable cash payout. The federal government retained over 200 million acres, including about 54 million acres in thirteen park units. This division largely settled the matter of royalties from oil, but it also raised the specter of federal land use regulation across a vast swath of the state, something both Natives and non-Native residents would come to resent. Unlike the Lower 48, national parks and preserves in Alaska are largely open to hunting, at least by Alaskan Natives. NPS policies about wilderness management, land access, and subsistence activities have been persistent sources of friction between the Service and state residents.[4]

Park units in Alaska also differ from those in the Lower 48 because they have limited access and visitation. Despite its enormous size, Wrangell-St. Elias has only about 200 miles of road, and only 65,000 people visited the park in 2022. Denali National Park, the best-known park in Alaska, has only 92 miles of road and 35 miles of developed trail across its 6 million acres. Peak visitation in 2018 was just below 600,000, and most of those visitors stayed on the park's main road. (In 2022, 1,127 people attempted the technical summit of Denali itself, and only 69% made it to the top.) As a result, national parks in Alaska do not have the same problem of overcrowding as occurs at parks like Yosemite or Acadia. In sum, the NPS's role in Alaska is unusual: it must manage vastly larger parks with a different set of rules than in the rest of the United States, and must do so in a highly fractious context.

The Mission of the National Park Service

Congress has given the NPS a mission that is clear on the surface but complicated to implement: preserve the national parks *and* facilitate their use. This broad mission requires constant negotiation as public and ecological pressures change. At its inception in 1916, the NPS placed a heavy emphasis on *increasing* visitation and use, and it succeeded wildly in the post-war period. Since then, high visitation and a growing public demand for ecological protection have necessitated more restrictive management.

The first two directors of the NPS were Stephen Mather and Horace Albright. The two were close friends and colleagues for decades, and together they pursued a vision of the parks that emphasized recreational opportunity on a large scale. Mather was a pragmatist. A successful businessman, he knew that the long-term success of the park system depended on a supportive political constituency. A broad base of support, in turn, required evangelization, marketing, and expanded visitor opportunities. Mather thought the masses needed to see the parks to be persuaded to support the Park Service. He and Albright worked closely with first the railroads and later the automobile associations to stimulate tourist interest. They endorsed the construction of road networks and grand lodges. They honed marketing campaigns—like *See America First!*, a slogan intended to win wealthy tourists away from Europe—to elevate park status among those who could afford to travel. Mather and Albright were true conservationists in their day, but their conservation challenges were quite different from ours today. They charted a course for the NPS that emphasized popular appeal—a course that is still evident now, reflected in the Service's ongoing concern for its positive image and its high rates of park visitation.

But this approach poses a vexing missional question. What *kind* of tourism and recreation should the parks allow? On this point, legal scholar Joseph Sax took issue with the Park Service's historic emphasis on mass tourism. In his acclaimed work *Mountains Without Handrails*, Sax argued that park management should prioritize contemplative recreation. "The parks are places that have not been tamed," Sax wrote, emphasizing nature's "peculiar power to stimulate us to reflectiveness by its awesomeness and grandeur, its complexity, the unfamiliarity of untrammeled ecosystems to urban residents, and the absence of distractions."[5] How would this play out for a particular management issue—say, how much automotive access to allow within the parks?

> The hotly contested question of vehicle use in the parks … is not an issue of transportation, but of pace. Intensity of concentration on the natural scene and attentiveness to detail are simply less likely to occur at forty miles an hour. For this reason it is appropriate to discourage motorized travel. Such a policy would not militate against all road building in reserved parklands. We need reasonable access to the various areas of very large parks. And because reserved lands should be made enticing to as wide a spectrum of the public as possible, including newcomers who need a taste of the opportunities the land offers, it makes sense to have—as we do in many parks—a highway designed to provide an introduction for those who are deciding whether they want to come back for more.
>
> The purpose of reserving natural areas, however, is not to keep people in their cars but to lure them out …[6]

Sax's philosophy, like any that favors preservation over easy access, creates additional difficulties for disadvantaged groups. And today, as at their origin, park policies are commonly evaluated against an equity criterion. At its inception, the American park system was hailed as an improvement upon the European aristocratic model, in which spectacular sights were enjoyed only by the wealthy. But parks that are difficult to explore will necessarily favor those with the time and the means to explore them. There is a deep tension here: the more we value park system lands for their wild and untamed qualities, the more difficult it becomes to keep those qualities within the reach of both rich and poor.

Rather than choose a single visitor experience in national parks, the NPS today works to provide a variety of experiences, from roadside tourism to wilderness backpacking. Most visitors want to be able to drive through a park, stop at a visitor center, find lodging, and buy food. Most large national parks have the infrastructure to meet these demands, albeit for a limited number of visitors. Congress and the NPS have also chosen to keep areas of large national parks in a state of limited development. Congress has created official wilderness areas in many national parks, and the NPS has largely avoided building new infrastructure outside of already-touristed areas. This approach aims to meet the demands not only of mass tourism but also of visitors who want an intensive, backcountry experience.

The Law of the National Park Service

The National Park Service is a large federal agency within the Department of the Interior. While NPS managers respond directly to visitor demands and political pressures, all agency decisions must ultimately be grounded in legal authority granted by Congress and directed by the president. The NPS is subject to all the general laws discussed in Chapter Four—the National Environmental Policy Act (NEPA), the Endangered Species Act, and so forth—as well as specific laws governing the national park system. The latter can be divided into three categories: the Park Service's Organic Act and its amendments; the legislation that created each of the park system's units; and the Service's own regulations, rules, and policies. This section will explore these categories.

The Organic Act

The Park Service gets its mission and legal mandate from the Park Service Organic Act of 1916, which created the agency, and from a handful of amendments and additions imposed by Congress in the years since. The Organic Act declares that the "fundamental purpose" of the Park System is:

> to provide for the enjoyment of the scenery, natural and historic objects, and wild life in such manner and by such means as will leave them unimpaired for the enjoyment of future generations.[7]

Flag this page for later because you should compare this statement of purpose with those of the agencies discussed in the coming chapters. Relative to the USFS and the BLM, the Park Service has a "dominant use" mandate: the job of the NPS is to manage its lands for present and future enjoyment by visitors. You'll notice a clear difference between this mandate and the "multiple-use" mandate of the USFS and BLM.

The NPS's majestic statement of purpose contains an obvious tension between *enjoyment* and *nonimpairment*. What kinds of enjoyment must be allowed and accommodated? What degree of impairment is too much? Let's consider both of these questions. Enjoyment means different things to different people. Some visit the parks for quiet seclusion. Others prefer to use the parks for motorized sports or thrill-seeking. Still others would love nothing more than to play a round of golf, take a gondola ride, or lounge at a luxury resort while gaping at natural wonders. Obviously, these activities are irreconcilable: the wilderness backpacker's experience would be diminished by loud ATVs, helicopter flights, or even large crowds, but restrictions on vehicles and crowds equally inhibit other kinds of enjoyment.

The nonimpairment mandate also bears on what activities may be allowed within the parks. Taken to its extreme, the idea of nonimpairment could prohibit almost all park access; even building a road within a park could be considered an impairment of sorts. But the Organic Act requires the Service to "provide for the enjoyment" of the parks, so clearly Congress did not intend for the Service to interpret nonimpairment in such a stringent sense.

So who decides which activities are allowed and what rules apply? Congress has the ultimate authority to do so, but it has, implicitly and explicitly, delegated much of that authority to the NPS. In other words, Congress can and sometimes does pass legislation that makes very specific managerial decisions, but much more often it simply provides broad guidelines, allowing the agency to make specific rules for specific parks. Generally speaking, the Organic Act gives the Park Service a great deal of leeway. As long as the NPS's decisions are careful and reasoned, they are likely to be upheld if a lawsuit is brought.

If a legal conflict arises between preservation of the park's resources and recreational activities or enjoyment, priority is usually given to preservation. For example, in a dispute about snowmobiling at Yellowstone, an appellate court once ordered the Park Service to reduce snowmobile activity within the park based on evidence that the level of activity previously allowed would disturb wildlife and impair the park's "natural soundscape."[8] Still, the general point remains: under the Organic Act, the Service has a great deal of leeway; few courts have been willing to overturn Park Service decisions about how to manage a particular park.

Park-Specific Legislation

As noted above, Congress sometimes writes specific laws pertaining to specific park system units. It might do this at the very time it is creating that unit, in the so-called park "establishment acts." Legislation establishing park system units often includes precise requirements (such as to build a specific road, or to prohibit the building of roads in a particular area) or makes exceptions to general park system rules (for example, to allow fishing and hunting or the landing of aircraft, all of which are generally disallowed within the park system). To get a sense for this sort of legislation, simply search the internet for "U.S. Code, Title 16, Chapter 1," and you'll be treated to the long list of laws that created all the various park units. Browse through them and you'll see that Congress is sometimes quite particular about rules for specific park units and other times is not.

BOX 5.1 An Excerpt from Title 16, Chapter 1, of the U.S. Code

Here is an excerpt, taken more or less at random from Chapter 1 of Title 16. It's part of the legislation establishing Indiana Dunes National Park. Notice the level of detail, and remember that this is an actual law, passed by Congress and signed by the President. How do you suppose such specific instructions came to be included in this legislation?

16 U.S.C. § 460u–15 – Rights-of-way; public access to Little Calumet River

> The Secretary may acquire only such interest in the right-of-way designated "Crossing A" on map numbered 626–91007 as he determines to be necessary to assure public access to the banks of the Little Calumet River within fifty feet north and south of the centerline of said river. The Secretary may acquire only such interest in the rights-of-way designated "Crossing B" and "Crossing C" on the map dated October 1986 and numbered 626–80,033–B as he determines to be necessary to assure public access to the banks of the Little Calumet River and the banks of Salt Creek within fifty feet on either side of the centerline of said river and creek.

16 U.S.C. § 460u–16 – Road construction cooperative agreements with landowners north of Little Calumet River; prevention of soil erosion; minimization of aural and visual impact

The Secretary shall enter into a cooperative agreement with the landowner of those lands north of the Little Calumet River between the Penn Central Railroad bridge within area II–E and "Crossing A" within area IV–C on the map referred to in section 460u–3 of this title, dated October 1976, and numbered 626–9100. Such agreement shall provide that any roadway constructed by the landowner south of United States Route 12 within such vicinity shall include grading, landscaping, and plantings of vegetation designed to prevent soil erosion and to minimize the aural and visual impacts of said construction, and of traffic on such roadway, as perceived from the Little Calumet River.

Does Congress ever interfere after a unit is created? The answer is yes. Our friend and colleague John Nagle offered examples of such instances:

> One statutory provision permitted the continued operation of a Kiwanis Club within Olympic National Park. Congress has also enacted numerous statutes governing scenic overflights at the Grand Canyon National Park. In another law, Congress directed NPS to "immediately cease" a plan that resulted in a settlement agreement to exterminate deer and elk on Santa Rosa Island in Channel Islands National Park. And Congress has repeatedly intervened to dictate the permissible snowmobile usage within Yellowstone National Park …[9]

Nagle also noted that Congress sometimes writes laws to overturn court decisions that it doesn't like:

> Numerous such statutory provisions have responded to judicial decisions contrary to the desired congressional management policy, including cases involving vehicle access across wilderness lands in the Cumberland Island National Seashore, the construction of a new bridge across the St. Croix National Scenic River, the exclusion of a renovated fire tower in Olympic National Park, the permissible cruise ship traffic in Glacier Bay National Park, and the continued operation of an oyster farm in Point Reyes National Seashore.[10]

That last excerpt reminds us that parks are political creations. Each park act is the product of negotiations and compromise, so it shouldn't be surprising that the mandates for different park units vary. Nor is it surprising that Congress, prompted by state delegations or perhaps the party in power, would periodically revise mandates for different park units. Congress has the power to require or prohibit certain activities within a specific park at any time, and sometimes merely the hint of congressional pressure is enough to influence Park Service decisionmakers. Local members of Congress, for example, became involved in disputes about snowmobiling in Yellowstone, airplane tours at the Grand Canyon, and off-road vehicles at Canyonlands. In each case, park officials held lengthy discussions with legislators before making a final decision about park rules. So congressional influence matters even when Congress does not formally legislate.

Park Service Regulations

Congressional statutes rarely provide sufficient detail to direct the work of day-to-day management, so Congress authorizes federal agencies to make additional rules by using the regulatory process. The very first statute to create a national park, the 1872 Act Establishing Yellowstone National Park, explicitly made it the duty of the Interior Secretary, "as soon as practicable, to make and publish such rules and regulations as he may deem necessary or proper for the care and management" of the park. Numerous laws since 1872 authorize yet more regulations. As noted in Chapter Four of this book, regulations are issued only after a thorough, structured process. Once issued, these regulations have binding force and legal effect for both the issuing agency and the general public. (Bear in mind, though, that regulations can be struck down by courts if they exceed an agency's authority or otherwise violate an act of Congress.) All the current NPS regulations are compiled in Chapter 1 of Title 36 of the Code of Federal Regulations, which you can easily find and peruse on the internet. Many regulations apply to all or to large portions of the park system; other regulations are limited to particular units. Box 5.2 provides an example of the specific regulations applicable to one park system unit, Hawaii Volcanoes.

BOX 5.2 The Regulations Pertaining to Hawaii Volcanoes National Park

Here are the federal regulations for one particular unit of the park system, Hawaii Volcanoes National Park:

36 CFR § 7.25 - Hawaii Volcanoes National Park.

a Fishing -

1 Commercial fishing. Commercial fishing from parklands (above the high waterline) other than as provided for below is prohibited.
2 Nets. The use of nets in fishing from parklands (above the high waterline) except for throw nets, is prohibited.
3 Kalapana extension area; special fishing privileges.

i Pursuant to the act of June 20, 1938, Native Hawaiian residents of the villages adjacent to the Kalapana extension area added to the park by the above act and visitors under their guidance are granted the exclusive privileges of fishing or gathering seafood from parklands (above the high waterline) along the coastline of such extension area. These persons may engage in commercial fishing under proper State permit.
ii For the purposes of this section, the term "native Hawaiian" means any descendent of not less than one-half part of the blood of the races inhabiting the Hawaiian Islands previous to 1778.

b Backcountry registration. No person shall explore or climb about the lava tubes or pit craters in the park without first registering with the superintendent and indicating the approximate length of time involved in the exploration and the number of people in the party. This section does not apply to the maintained trail through Thurston Lava Tube, nor the maintained trail down and across Kilauea Iki pit crater.

There are other ways that agencies can make rules, even if they lack the formality found in the Code of Federal Regulations. In fact, there is a range of tools available. For example, the Park Service distributes among its employees a document called "Management Policies." (You can find this on the internet, too.) This document contains over 150 pages and covers everything from soundscape management and overflights to fireworks and merchandise sales. These policies are intended to serve as a unified, consistent set of internal operating instructions for NPS workers. Even so, local park officials sometimes have substantial "wiggle room" in how they manage their unit.

To summarize the last few pages, directive legislation, regulation, and NPS policies together establish the rules that bind and guide the Park Service. In any given visit, a tourist might encounter the following rules without knowing whether they emanate from Congress, from the NPS Headquarters, or from that particular park's administrator:

- No backcountry camping except by permit, and none within 100 feet of a water source.
- No parking along this road between 10 p.m. and 6 a.m.
- Hang-gliders must pre-register at the Administrative Headquarters.
- Bicycles must remain on paved pathways.

But there is one final, powerful tool that we have not yet mentioned: the federal budget. After all, the Park Service can only work when it has adequate money with which to procure staff, contractors, and materials. Sometimes, Congress uses its annual appropriations bills to explicitly prohibit the agency from spending money on certain initiatives. At other times, it does the opposite, appropriating money for specific projects. For example, the NPS recently estimated that it had a $20 billion maintenance backlog on roads, buildings, and other infrastructure within the park system. To address this, Congress in 2020 passed the Great American Outdoors Act, which provided $1.3 billion per year for 5 years to address this backlog, plus another $900 million annually for the NPS to purchase land and fund state initiatives in conservation and recreation. Budgetary decisions like this can significantly affect the agency's work.

The National Park Service Operations

So how does the National Park Service implement these statutes and regulations across 50 states and hundreds of individual properties? After all, it is a massive agency, with over 20,000 full-time and seasonal employees and a budget of over $3 billion.[11] To understand how it all works, let's look at the agency's organizational structure (Figure 5.2).

The NPS is led by a director who is nominated by the sitting president and confirmed by Congress, which often causes priorities to shift from one administration to the next. The agency's functions are then split into three areas, overseen by deputy directors: operations, congressional and external relations, and administration. The deputy director for operations leads the management of all park units. Operations is further divided into programmatic areas that reflect the agency's work priorities. For example, one associate director provides guidance for "interpretation, education, and volunteers" across all park units. These associate directors provide general guidance that ensures consistency throughout the agency. The deputy director for operations also oversees twelve regional directors who, in turn, supervise all the park units in their geographical area. Park units themselves range widely in staff number and expertise.

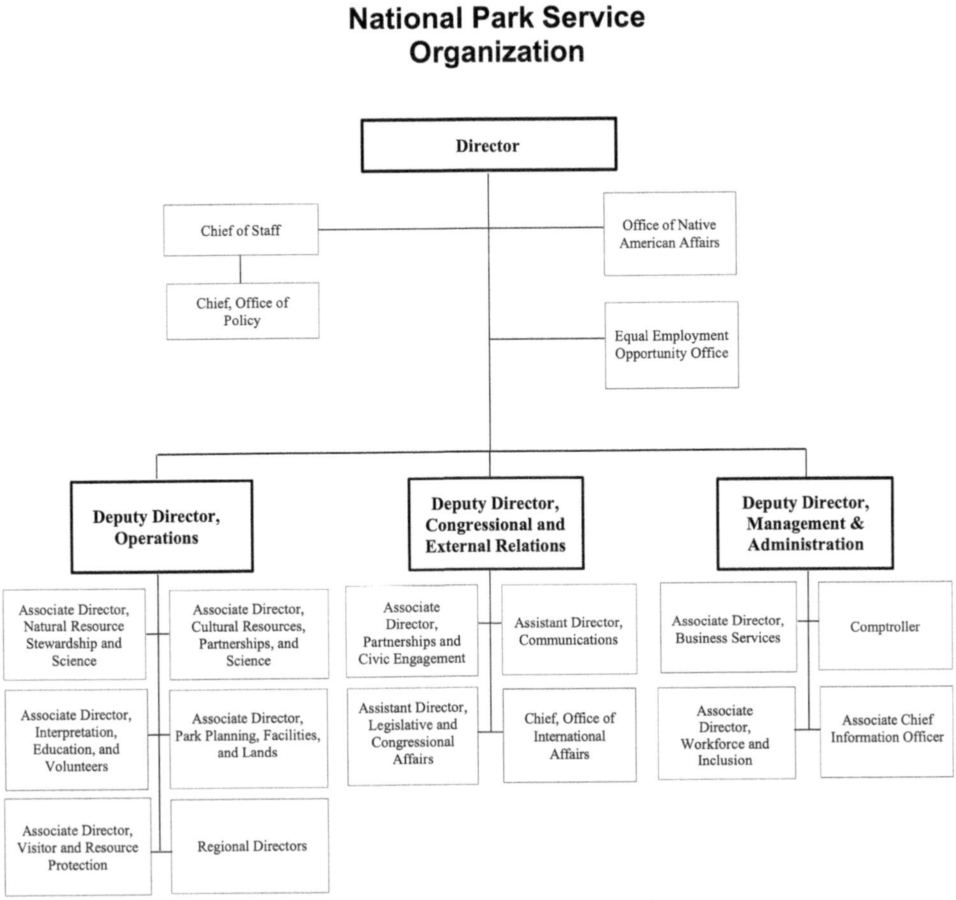

Figure 5.2 Organizational Chart of the National Park Service. Source: www.nps.gov.

For comparison, Yellowstone National Park has historically employed a permanent staff of nearly 800, while Kenai Fjords National Park in southern Alaska has just 30.

Operations is only one of three deputy directorates. A second deputy director oversees external relationships, managing NPS communications and the agency's relationships with federal and state lawmakers and agencies. The third deputy director oversees administration of this complex organization, which includes things like budgeting and fee collection, IT services, and human resources.

For those interested in a career with the agency, seasonal employment is an excellent way to start. Of the agency's roughly 20,000 employees, almost 40% are seasonal. While many employees return year after year, the NPS still has thousands of openings every year. And it may be surprising to learn that NPS employees make up only half of all the people working in national park units. That is because the NPS outsources most concessions—such as food, lodging, guide services, and transportation—to accommodate the growing number of visitors. In 1916, when the NPS was created, an estimated 326,506 people visited the national parks. By 2022, the number of visitors had increased by two orders of magnitude to 311,985,998. Today, there are about

500 permitted concessionaires in the national park system, employing 25,000 people (mostly seasonal workers) and generating almost $1.5 billion in revenue.[12]

Conclusion

The National Park Service plays a crucial role in stewarding some of the most famous tourist sites in the world. Its mission to preserve and interpret natural and cultural sites while promoting their use and enjoyment means that almost no commercial resource development—logging, mining, grazing—occurs inside park units. Commercial development is limited to meeting the comparatively inobtrusive needs of its concessionaires. The agency still struggles with the inherent tension in its mandate between preservation and use, but the Park System remains for many the most recognizable, accessible, and tourist-friendly of the public land systems. Of the four primary federal land agencies, the NPS is best situated to welcome the full range of visitors. It manages wilderness areas, where visitors must enter on foot or animal; it manages campgrounds for auto tourists interested in a rustic camp experience; and it leases hotels and restaurants to concessionaires to serve those who want more modern comforts. As you read about the other three agencies in the coming chapters, compare and contrast their missions and challenges with what you've learned in this chapter.

Resources for Further Study

Albright, Horace M., and Robert Cahn. *The Birth of the National Park Service: The Founding Years, 1913–33* (Salt Lake City, UT: Howe Brothers, 1985).

Keiter, Robert B. *To Conserve Unimpaired: The Evolution of the National Park Idea* (Washington, D.C.: Island Press, 2013).

Nagle, John Copeland. "How National Park Law Really Works." *University of Colorado Law Review* 86 (2015): 861–926.

Runte, Alfred. *National Parks: The American Experience* (Plymouth, UK: Taylor Trade Publishing, 4th edn. 2010).

Sax, Joseph L. *Mountains Without Handrails: Reflections on the National Parks* (Ann Arbor, MI: The University of Michigan Press, 1980).

Spence, Mark David. *Dispossessing the Wilderness: Indian Removal and the Making of the National Parks* (New York, NY: Oxford University Press, 1999).

Notes

1 There are another 150 related areas outside of the formal National Park System for which the National Park Service bears more limited responsibility.
2 Executive Order No. 14253, "Restoring Truth and Sanity to American History," Mar. 27, 2025.
3 Sometimes, the NPS has difficulty evicting these private parties even when their property rights have expired. See Bruce R. Huber, "The Durability of Private Claims to Public Property," *The Georgetown Law Journal* 102 (2014): 991–1043, 1015 ff.
4 See Andee Sears, Adrienne Lindholm, and Peter Christian, "ANILCA: A Perspective From Boots on the Ground," *Alaska Park Science* 21 (2022): 34–51.
5 Joseph Sax, *Mountains Without Handrails: Reflections on the National Parks* (Ann Arbor, MI: The University of Michigan Press, 1980), 111, 46.
6 Sax, *Mountains Without Handrails*, 79.
7 54 U.S.C. § 100101(a).
8 Robert Keiter, *To Conserve Unimpaired: The Evolution of the National Park Idea* (Washington, D.C.: Island Press, 2013), 80.

9 John Copeland Nagle, "How National Park Law Really Works," *University of Colorado Law Review* 86 (2015): 861–926, 909.
10 Nagle, "How National Park Law Really Works," 910.
11 Congressional Research Service, "National Park Service: FY2022 Appropriations." Please note that these numbers are always changing, thanks to the federal budget process, changing presidential administrations, etc. The second Trump Administration, for example, notoriously trimmed the size of the entire federal workforce shortly after taking the reins from the Biden Administration.
12 These statistics were pulled from the nps.gov website in mid-2025.

6 The National Wildlife Refuge System

Introduction

The National Wildlife Refuge System is strikingly different from the National Park System. Wildlife refuges are not flashy tourist sites; many Americans would be unable to name a single National Wildlife Refuge. They tend to offer few amenities, and they are seldom chosen as a vacation destination. But they are not without a constituency. Hunters, birders, scientists, and other wildlife enthusiasts are tireless supporters of the refuge system. This chapter begins with an overview of the lands of the refuge system, then discusses the mission, guiding laws, and organization of the U.S. Fish and Wildlife Service (FWS) and the refuge system it stewards.

Before we dig in, a few explanatory points are in order. First, the FWS isn't solely or even primarily a land management agency. Congress has assigned various conservation-related tasks to the FWS, including such weighty matters as designating and rehabilitating endangered and threatened species. This chapter will briefly describe these tasks, but because this is a book about the public lands, it will focus primarily on the FWS's management of the refuge system itself.

The second point has to do with the word "refuge." Some regard it as a misnomer because hunting is often allowed on refuge system lands. In fact, within many wildlife refuges, hunting is the most prominent recreational activity. For the critters being hunted, they are not much of a refuge! Yet sport hunters are reliable advocates for wildlife preservation in America, and through the years their purchases of "Duck Stamps"—a collectible that also serves as a federal hunting license—have paid for many additions to the system. Since its earliest days, the refuge system has been championed both by purists who wouldn't dream of wielding a weapon *and* by hunter-conservationists in the mold of Teddy Roosevelt. Its appeal to these disparate groups helped the refuge system grow into the expansive system it is today. With those points in place, let's examine the properties within the refuge system.

The Lands of the National Wildlife Refuge System

Like all the public land systems, the National Wildlife Refuge System encompasses a wildly diverse array of lands. But while there is no "typical" national park, there *is* a typical wildlife refuge. The standard wildlife refuge is a modest area, rich in wetland habitat for waterfowl, and located within one of the four major flyways that run from north to south across North America. Wildlife refuges dot these flyways like rest areas on an interstate freeway, providing seasonal living quarters for millions upon millions

DOI: 10.4324/9781003392514-6

of migratory birds. The scenery at refuges, though pastoral and often beautiful, tends toward the mundane.

The wildlife refuge system contains habitat for many kinds of animals, but birds are first among equals. If you're not a bird lover, you might seldom notice or think about these flying wonders. Birds have long attracted loyal observers ("birders"), many of whom are willing to wait patiently in remote landscapes, binoculars in hand, for a glimpse of a favorite species. More troubling, rare and exotic birds make valuable targets, both for culinary delicacies (like hummingbirds' tongues) and illicit trade. In the early twentieth century, exotic feathers were sought in large quantities for use in plumed women's hats (see Figure 6.1). Before long, birders noticed that certain species were in sharp decline.

The ardent naturalist President Teddy Roosevelt played a prominent role in abating the slaughter of birds for the fashion market. As described in Chapter Three, Roosevelt strained his presidential powers to create the first wildlife refuge, Pelican Island, in southeast Florida, after it became clear that the enormous birds were in jeopardy due to the sudden and rapacious market for feathered hats. By the time Roosevelt's presidency ended, he had designated more than fifty refuges, most of them also positioned to protect known habitat for overhunted bird species. These were public lands where hunters would otherwise have been able to hunt without restriction. Congress soon joined in, ratifying Roosevelt's designations and adding even more units to the system. Later presidents continued the trend, and now there are nearly 600 units within the National Refuge System.

Figure 6.1 Actress Lillian Russell wearing a plumed hat, typical of early-twentieth-century fashion. Source: Library of Congress.

Roosevelt, a lover of wild things, was himself an avid hunter. To him, this was no contradiction. What was true of Roosevelt as an individual was also true of America at large: Americans sought to preserve wildlife, but they also loved to shoot it—for food, for entertainment, and for management of their natural surroundings. While America's "gun culture" is now a hot-button political concern, one hundred years ago it was simply a fact of life that rural Americans owned and used their guns. It would have been unusual to live in the countryside without one. The need to shoot predators to protect family and livestock was simply assumed. Similarly, businesses dependent on harvested animals were no more morally questionable than the consumption of meat for food. So perhaps it is no surprise that the wildlife refuge idea began to evolve the moment it was established. Soon, refuges became places where *some* hunting was protected, though limited to particular species and seasons.

The paragraphs that follow will discuss characteristics of the lands in the refuge system. By the end, you should have a feel for the system's composition. But bear in mind that, as with all the public land systems, there are outliers that defy generalization. This is especially true of the refuge system. A leading expert, Robert Fischman, once described the system as "a crazy quilt," "a tangle of land units with widely varying sizes, purposes, origins, ecosystems, climates, levels of development and use, and degrees of federal ownership and Service control."[1]

Flyways

Ornithology came into its own as a science in the early 1900s, and one of its first lines of inquiry had to do with bird migration. It was evident that certain birds moved seasonally; what was not yet known was where those birds went—how far they traveled; what routes they used; whether these variables changed or remained the same; and so forth. New modes of travel and study eventually allowed scientists to track bird populations with greater precision. Bands were attached to birds' legs, enabling specific animals to be observed and identified at multiple locations of their journey. What scientists discovered was stunning. Migratory routes were far longer and more consistent than had yet been known. By roughly the 1920s, ornithologists could label four distinct "flyways"—broad migratory paths traveled not by one but by myriad species.

Wildlife refuges are heavily concentrated within these four flyways (Pacific, Central, Mississippi, and Atlantic), with perhaps three-fourths of the refuges set along them. Growing familiarity with the flyways also helped establish an agenda for the addition of new refuges: wildlife specialists could identify needed habitat along a migratory path and look for viable properties. Nowadays, hunting communities understand that wetland conditions a thousand miles away can affect their local hunting season.

Rivers, Coasts, and Wetlands

Along a flyway, what sort of land is most suitable for a wildlife refuge? If you've ever paid attention to flocks of birds, you may have noticed that their movements depend on access to water. Flocks will seldom come to rest anywhere without significant water, whether a pond, river, swamp, or estuary. Even standing water on a field can provide a temporary home. The refuge system reflects the importance of wetland habitat to waterfowl; the majority of refuges lie along rivers, shorelines, and other wetland spaces.

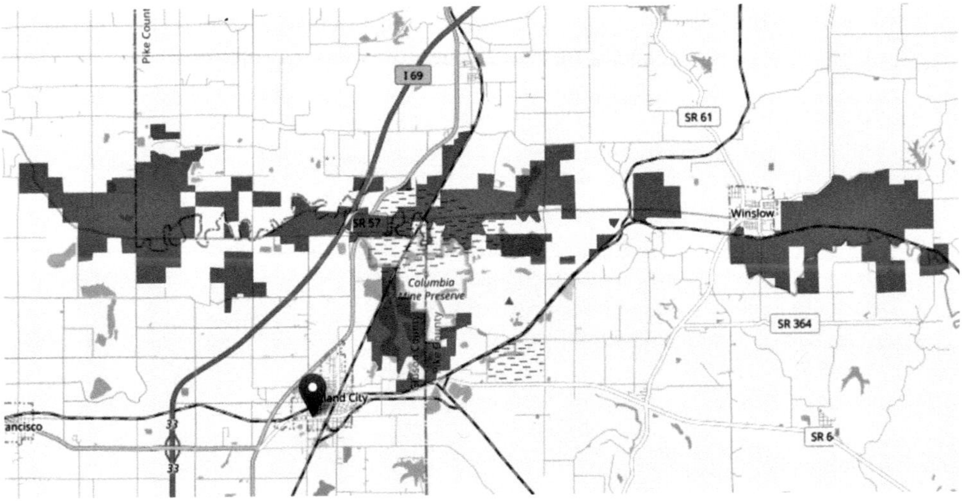

Figure 6.2 The dark grey areas make up the Patoka River NWR in Southern Indiana. Note the discontinuity of the lands comprising the refuge. This is not particularly unusual, especially for refuges in the eastern U.S. Source: OpenStreetMap.

This fact affects the shape, size, and composition of the refuges. You may recall from earlier chapters that settlers often chose land adjacent to water. Wetlands and lands bordering waterways were the first to be claimed and privatized. So as the refuge system grew, many wetlands critical for wildlife protection would have to be purchased from landowners rather than reserved from the public domain. Federal officials, always fiscally constrained, were tasked with acquiring private land but were compelled to select the most limited acreage sufficient to accomplish federal purposes. This is one reason why many wildlife refuges are quite small. Many are cobbled together out of several parcels, in sharp contrast with large national parks, forests, or monuments, which were simply carved out of the expanse of public lands that had never fallen into private ownership. Today, hundreds of national wildlife refuges are under 10,000 acres in size (this is roughly the amount of land in a square with four miles to a side), and many are fragmented among discontiguous wetland parcels (see Figure 6.2 for an example). This fragmentation would seem quite odd if the principal purpose of the refuge system was to appeal to tourists and recreationists. But the objective is to protect habitat for waterfowl, and from that standpoint the assembly of nonadjacent parcels makes more sense.

BOX 6.1 Wildlife Refuges for Non-Bird Species

Birds are obviously not the only form of protected wildlife, and numerous wildlife refuges provide habitat for other species. The Endangered Species Act specifically authorizes the FWS to acquire lands for the conservation of endangered or threatened species. For non-bird species, wetlands may be less critical, and other habitat characteristics come to the fore. For example, the National Bison Range in Montana and the National Elk Refuge in Wyoming were selected out of mountainous territory largely untouched by settlement but rich in summer forage.

The list below provides a sampling of units in the National Wildlife Refuge System which were not chosen primarily for their habitat for birds:

- National Elk Refuge
- National Bison Range
- Wyoming Toad Conservation Area
- Hart Mountain National Antelope Refuge
- Julia Butler Hansen Refuge for the Columbian White-Tailed Deer
- Ozark Cavefish National Wildlife Refuge
- Northern Tallgrass Prairie National Wildlife Refuge
- Mountain Longleaf National Wildlife Refuge

Waterfowl Production Areas

One of the more unusual classifications within the refuge system is the Waterfowl Production Area (WPA). WPAs are a very specific type of protected land found almost entirely within a particular biophysical region called the Prairie Pothole Region. This region forms the northern part of the Central Flyway, stretching across the Great Plains from Northern Alberta to Iowa. The potholes themselves are shallow depressions, formed by receding glaciers, that fill with water from rain and melting snow. By late summer, many of these potholes are dry, although some retain water year-round.

Given their shallow water and surrounding grasslands, and their felicitous position on a major flyway, Prairie Potholes are prime habitat for various migratory waterfowl—so much so, in fact, that the Pothole Region is sometimes called "America's Duck Factory." Despite their importance, the potholes remained legally unprotected at midcentury. The lands of the region had already passed out of the public domain, and many potholes were drained or otherwise damaged by farmers, who wished, understandably, to devote their land to productive use during the growing season.

This context produced an interesting legal accommodation. When the federal government sought to protect the potholes, it recognized the need to minimize disruption to the agricultural activity around them. Rather than purchase land outright, the government employed a legal device called the "conservation easement." Under such an easement, a landowner is paid *not* to use the potholes for any purposes that would diminish their habitat viability. For example, a landowner who sells a grassland easement to the Service might agree to maintain permanent vegetative cover on the land subject to the easement, while retaining the right to mow, hay, and seed the land after July 15 of each year. Another landowner might agree not to drain, fill, or level the relevant wetland, while retaining the right to farm, graze, or hay any land that dries up naturally. Conservation easements provide no public right to recreational use. So by using easements, the FWS is essentially able to delegate wildlife management to individual landowners, saving the time and money that would be required to own and operate a full-fledged wildlife refuge. It's for these reasons that out of the 4 million acres contained within WPAs, roughly 3 million are in the form of conservation easements. This underscores a key difference between the National Wildlife Refuge System and the National Park System: the refuge system does not exist primarily to serve tourists or to provide recreational opportunity but to protect wildlife habitat.

Figure 6.3 The Prairie Pothole Region of North America. Source: Wikimedia Commons.

Alaska

Yet again, Alaska stands alone. Alaska's wildlife refuge acreage vastly exceeds that of all the other states put together. Over 76 million of the roughly 95 million acres of land in the refuge system are in Alaska, including the eleven largest refuge units. The biggest, the Arctic National Wildlife Refuge (ANWR), is larger than the state of South Carolina. ANWR is renowned for more than just its size. Sitting adjacent to Prudhoe Bay, it overlies truly massive oil fields but is also home to Indigenous communities and to incredible ecological diversity, from polar bears to caribou to shorebirds. For

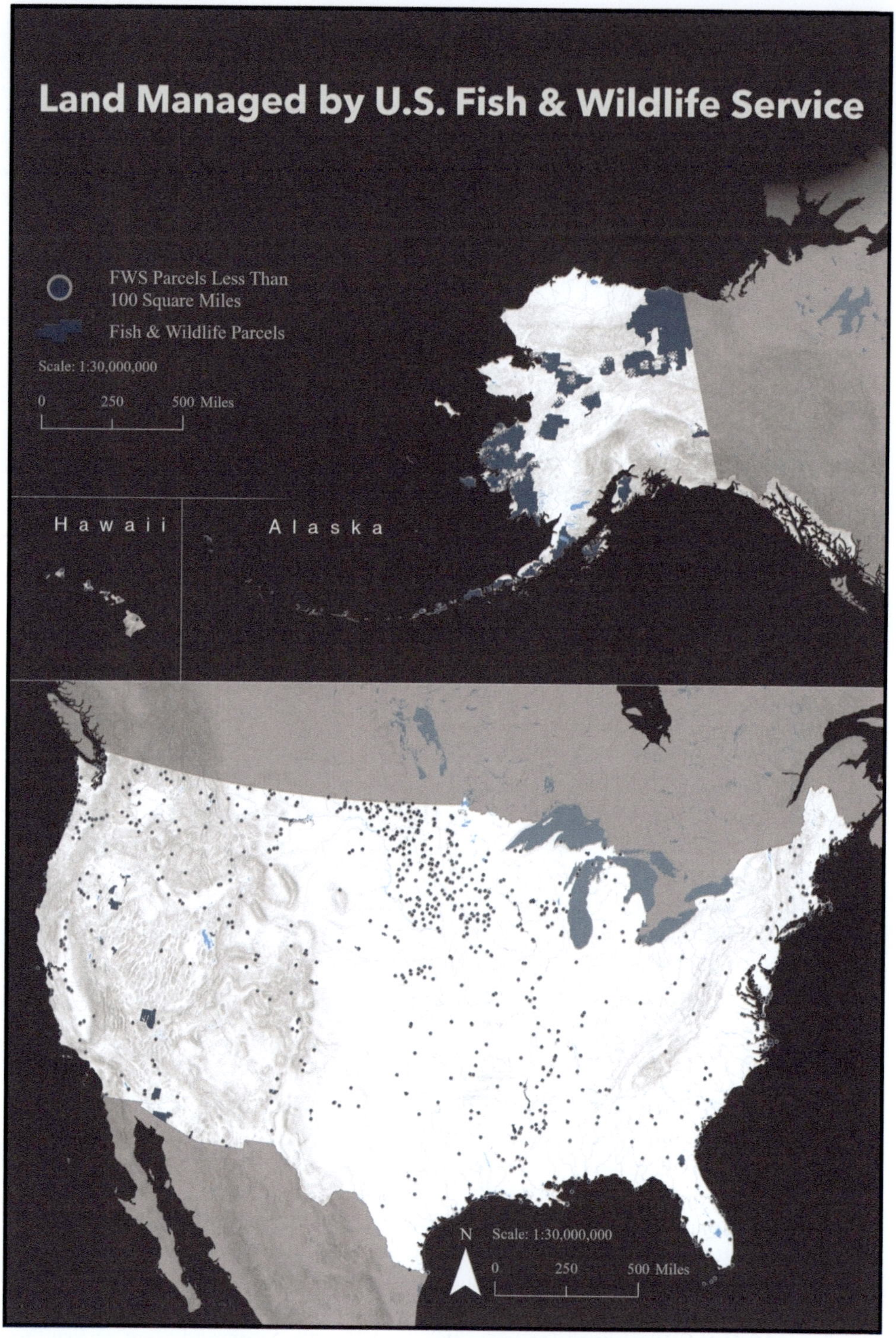

Figure 6.4 Map of National Wildlife Refuges. Note the clustering of refuges in the Prairie Pothole Region and the size of the Alaskan refuges. Map by Addison Arendt.

decades, politicians fought savagely over whether to open ANWR to oil drilling. The high-profile issue became standard fodder even among presidential candidates. In recent years, the dispute has subsided somewhat. During the first Trump Administration, Congress passed legislation opening ANWR for oil and gas leasing, but the industry showed little interest, perhaps recognizing that the political risk associated with drilling in such fragile circumstances had not disappeared. And indeed, the Biden Administration wasted little time in canceling the few leases that had been issued during Trump's first presidency.

Adding Lands, Changing Boundaries

Like the other public land systems, the National Wildlife Refuge System is not static. Each year, the FWS buys, sells, and swaps land. For many years, the FWS was unique among the four land management agencies in that it alone enjoyed annual, guaranteed federal appropriations for land acquisition. This is because federal law requires Duck Stamp revenue to be made available to the Service for the "location, ascertainment, and acquisition of suitable areas for migratory bird refuges."[2] Now, the Great American Outdoors Act of 2019 has authorized mandatory annual appropriations for land acquisition in other public land systems also.

Because the FWS has been engaged in a slow but steady campaign of acquisition, it has developed a detailed process for identifying and prioritizing target properties. Even after that process is complete, the law may require consultation with other state or federal officials before a transaction can be consummated. The FWS also abides by two important policies in acquiring land: first, it purchases only the "minimum interest necessary" to accomplish system objectives (e.g., if possible the agency will purchase an easement rather than full title); and second, the agency buys land from "willing sellers only." Eminent domain, for the time being, is off the table. This creates situations in which Congress has authorized the creation or enlargement of a wildlife refuge, yet the landowners within that area remain unwilling to sell (or FWS without sufficient funding to buy). For example, Cache River NWR in Arkansas has an approved acquisition boundary of 285,937 acres, but in its present state consists of only 72,606 acres. Over time, depending on its other priorities and on landowners' willingness to sell, the FWS will continue, slowly but surely, to add land to the Cache River NWR.

The Mission of the National Refuge System

Like the National Park System discussed in Chapter Five, the National Wildlife Refuge System is a dominant use system. This means that the FWS is required to prioritize a specific objective (whereas the Forest Service and the BLM must balance multiple uses, as you'll see in the next two chapters). Federal law dating to 1997 assigns FWS the following mission:

> to administer a national network of lands and waters for the conservation, management, and where appropriate, restoration of the fish, wildlife, and plant resources and their habitats within the United States for the benefit of present and future generations of Americans.[3]

Perhaps because the law is relatively clear, or perhaps because the refuge system has low public visibility, the FWS has not experienced protracted philosophical struggles over how to manage its lands. Its leaders have not historically been public figures, and debates over its management have not attracted much public attention (with the very important exception of oil drilling in the Arctic NWR discussed above).

Nonetheless, there are important ongoing questions about how the FWS interprets and carries out its mission. As noted earlier, for example, the conservation-oriented mission of the agency sits somewhat uneasily with hunting, which has long been a priority use of wildlife refuges. Some argue that any apparent conflict between hunting and conservation is superficial because hunting can carry important ecosystem benefits. The FWS itself has declared that "in some instances [hunting] is necessary for sound wildlife management" because, for example, "deer populations will often grow too large for the refuge habitat to support."[4] But we should not be surprised when there is public outcry about hunting in the refuge system, as for example when the Trump Administration in 2020 authorized expanded hunting and fishing access across a large portion of the system. The hunting-versus-conservation conflict is essentially "baked in" to the FWS's culture and history.

Also, notice the reference in the FWS mission statement to "plant resources." As its history suggests, the refuge system was not initially created to protect plants; only with the 1997 legislation was floral conservation added to the FWS's mission. It is a significant expansion, one that tracks a broader shift among the land management agencies toward a model of ecosystem management.[5] Other provisions in the same legislation make clear that Congress meant what it said, requiring the FWS to monitor plant life and to include consideration of flora in its comprehensive planning process. Yet the Service has been slow to embrace the breadth of this expansion, choosing instead to emphasize wildlife despite the equal importance suggested by the phrase "fish, wildlife, and plant resources and their habitats" in the mission statement.[6] The phrase "wildlife first" still appears in some FWS publications and regulations. Although "wildlife first" certainly reflects the agency's origins and expertise, some would argue that the agency is missing an opportunity to take a broader view of its ecological mandate.

The Law of Wildlife Refuges

As discussed in Chapter Four, all the public land agencies, FWS included, must comply with federal environmental law: the National Environmental Policy Act, the Endangered Species Act, the Clean Water Act, and so forth. Beyond that, wildlife refuges are subject to both state and federal law, the latter itself a mix of site-specific and system-wide rules. Let's begin by discussing the federal legislation that created individual wildlife refuges.

Site-Specific Legislation

National parks and forests, at the time of their creation, became part of a recognizable system quite quickly. Although it took time for the NPS and USFS to establish themselves as major federal agencies, from an early date the units of those systems shared well-understood commonalities. Not so with the National Wildlife Refuge System. Congress created wildlife refuges in a more haphazard fashion, each new refuge bringing with it a specific history and a specific federal mandate. Congress didn't put the refuges together into a single system until the mid-1900s, and even that legislation received significant updates right up until 1997 with the passage of the National Wildlife Refuge System Improvement Act, which we'll examine in the next section.

Because of this history, the legislation authorizing or creating each individual refuge remains critically important. Refuges designated by executive orders and proclamations, Secretary's Orders, public land orders, and refuge-specific legislation take on the purposes described in the establishing document. If a refuge was established pursuant to a general conservation law, that refuge simply takes on the purposes of the law. For example, Lost Trail NWR in Montana, like many early refuges, was established under the authority of the Migratory Bird Conservation Act. Its purpose comes directly from that Act, which describes setting aside land "... for use as an inviolate sanctuary, or for any other management purpose, for migratory birds."[7]

Although Congress has now consolidated the refuges into a system and has passed legislation giving a coherent legal mandate to the FWS, Congress has never wiped the slate clean of the requirements imposed by the refuges' establishing documents. To the contrary, federal law is quite clear that individual refuge purposes trump the overall mission of the system:

> [I]f a conflict exists between the purposes of a refuge and the mission of the System, the conflict shall be resolved in a manner that first protects the purposes of the refuge, and, to the extent practicable, that also achieves the mission of the System.[8]

Yet precisely because of the diverse history among the various refuge units, determining a refuge's purposes is not as straightforward as it might appear. Many refuges have multiple purposes from a variety of authorities. Refuges may have been cobbled together over time by various actors, or the establishing document itself may list several purposes for the refuge. For example, Malheur NWR's purposes are found across a dozen congressional, presidential, and administrative documents.

Sometimes, the multiple purposes specified for a given refuge may conflict. For example, Crab Orchard NWR in southern Illinois is to be managed for industrial and agricultural use as well as the more typical recreational and wildlife conservation purposes.[9] Tule Lake NWR in northern California is "... dedicated to wildlife conservation ... for the major purpose of waterfowl management, but with full consideration to optimum agricultural use that is consistent therewith."[10] Thus, regardless of whether allowing agricultural use is consistent with the overall mission of the Refuge System, the refuge manager is still required to allow it on Tule Lake so long as it is consistent with proper waterfowl management. The agency's own Policy Manual provides another example:

> [A] hypothetical conflict between conservation of Key deer and other wildlife at National Key Deer Refuge would be resolved in favor of the Key deer because the [National Wildlife Refuge System] Improvement Act states we should manage each refuge to fulfill the specific purposes of that refuge (e.g., "to protect and preserve ... the Key deer and other wildlife resources in the Florida Keys").[11]

System-Wide Legislation

The foregoing detour into site-specific legislation should help you understand why the National Wildlife Refuge System sometimes seems a bit less coherent than the National Park System. When the national network of wildlife refuges was consolidated in 1966, it did not eliminate the diversity among the units. Congress has understandably sought to honor the initial purposes underlying each refuge's creation but by doing so has made it difficult to develop a comprehensive national approach to refuge management.

An enduring legal challenge for the Refuge System, then, lies in deciding what activities to allow within wildlife refuges. Congress ultimately helped answer that question by passing the National Wildlife Refuge System Improvement Act in 1997, which we'll simply call the "Improvement Act" from now on. What was the Improvement Act improving? It was improving the Refuge System by clarifying how exactly the FWS should determine which activities could occur on specific wildlife refuges. The impetus for the Improvement Act was a problem that had emerged in the decades prior. In 1989, an influential federal investigation noted that "refuge managers report that activities they consider harmful to wildlife resources (such as power boating and off-road vehicles) are occurring on nearly 60% of the wildlife refuges."[12] Specifically, the investigation found that so-called "secondary uses" of the refuges—non-wildlife-related uses such as public recreation and mining—were commonplace even though they were incompatible with the principal mission of the Refuge System. The primary reasons were found to be local public pressure and FWS's incomplete ownership and control over refuge system lands (because of, e.g., private mineral rights within a refuge or a landowner's control over a Waterfowl Production Area).

Congress sought to address these problems in the 1997 Improvement Act, which, according to law professor Robert Fischman, functions as an overarching "organic act" for the refuge system.[13] Fischman observed that, read carefully, the Improvement Act implies a hierarchy of allowable uses on the refuges. This five-layer hierarchy can be applied by refuge managers when determining whether to permit a proposed use on a refuge unit.

- As noted above, individual refuge purposes, and refuge management activities designed to accomplish them, remain the highest priority at all times. These activities might include water level management, invasive species control, scientific monitoring, and routine maintenance. While it is uncommon for these tasks to conflict with the Refuge System's overall mission of conservation, conflicts might occur in refuges designated specifically for the protection of a certain endangered species or ones that place a high priority on agricultural or mineral development.
- The next layer in the hierarchy is the mission of conservation—the sustenance, restoration, and enhancement of "healthy populations of fish, wildlife, and plants," which the Improvement Act facilitates by authorizing refuge management activities such as "protection, research, census, law enforcement, habitat management, propagation, live trapping and transplantation, and regulated taking."[14]
- Third in the hierarchy are wildlife-dependent recreational or educational uses such as hunting, fishing, wildlife observation and photography, and environmental interpretation. Once the FWS determines that such a use is compatible with a refuge's purposes and the overall mission of conservation, that use "should be facilitated, subject to such restrictions or regulations as may be necessary, reasonable, and appropriate."[15]

- Non-wildlife-dependent recreational uses, such as cross-country skiing or mountain biking, receive the next highest priority. These uses may be permitted on refuge units, but only to the extent that they are compatible with any wildlife-dependent uses, the refuge's purposes, and the overall System mission. If permitted, these uses must be reevaluated at least every 10 years.
- Finally, refuge management economic activity and other economic uses (such as logging, grazing, or mining) receive the lowest priority in the Refuge System. These uses are only permitted if they are compatible with wildlife-dependent recreational uses *and* contribute in some way to the refuge's purposes or the mission of the system. These uses, if approved, are also subject to a reevaluation every 10 years.

You may have noticed that the last three categories all rely heavily on the word "compatible," so it's important here to consider this word carefully. According to the FWS policy manual, a compatible use is

> [a] proposed or existing wildlife-dependent recreational use or any other use of a national wildlife refuge that, based on sound professional judgment, will not materially interfere with or detract from the fulfillment of the National Wildlife Refuge System mission or the purposes of the national wildlife refuge.[16]

Compatibility is thus the most significant criterion that refuge managers must consider before introducing new uses to refuge lands. Part and parcel of the compatibility determination is that, before allowing a newly proposed use of a refuge, or renewing an existing use, managers are required to allow and consider public comments about those uses and to consider fully their effects in light of the hierarchy of uses outlined above.

State Wildlife Law

Importantly, *state* wildlife law often applies to *federal* wildlife refuges. The best example of this has to do with the regulation of hunting seasons and catch limits. Hunters and fishers don't escape the restrictions of state rules simply by walking onto a federal reserve. In fact, anyone hunting or fishing on a national wildlife refuge must have a state license and remains subject to the seasonal schedule and bagging limits established by state law. Congress has specifically required the FWS to maintain consistency with state fish and wildlife laws and regulations to the extent practicable.

Nevertheless, the U.S. Constitution declares that when state and federal law collide, federal law is supreme. And they do collide from time to time. Here's an example. In the late 1990s, the state of Wyoming began vaccinating its wild elk against a disease called brucellosis, which had been found within the local elk population. Because brucellosis can spread from elk to cattle, its presence posed a danger to Wyoming's cattle industry. In fact, several states threatened to disallow the entry of Wyoming cattle to stem the spread of brucellosis. But when Wyoming sought to vaccinate the elk on the National Elk Refuge located inside the state, the FWS refused to allow it. The Service claimed the vaccine hadn't been adequately tested and proposed to launch a multi-year testing program. The Wyoming cattle industry couldn't afford to wait several years, though, so Wyoming sued the FWS, arguing that in this instance the agency was obliged to yield to the state's right to manage wildlife within its borders. The federal courts decided in favor of the FWS, stating that Congress intended the Service to have primary authority, and thus primary decision-making power, over wildlife located on National Wildlife Refuges.[17]

A similar situation unfolded more recently in Alaska, where bear hunting has long been restricted. In 2013, the state determined that the brown bear population in Kenai National Wildlife Refuge had become sufficiently large to justify loosening hunting restrictions on the animal. The Service, however, was concerned that the lack of restrictions, particularly the allowance of taking brown bears over bait traps, would lead to unsustainable harvest levels in the Kenai brown bear population. Because of these concerns, the Service blocked the state's authorization of brown bear baiting in the refuge.[18] The state of Alaska sued the Service, arguing that the agency was unlawfully interfering with state wildlife management. Again, however, the federal courts rejected the claim.[19] State hunting laws, the court declared, are ultimately subject to any regulations imposed by the Service. You can be quite sure, however, that this won't be the end of the story; Alaska and the federal government are constantly engaged in disputes over the management of federal lands within the state.

Managing the Refuge System: The Organization and Operations of the FWS

At the outset of this chapter, we noted that the FWS is not solely a land management agency. It's time now to explain a bit more about this important federal entity. Over the years, Congress has assigned many important programs to the FWS or its predecessors. It is one of two federal agencies tasked with administering the Endangered Species Act. It runs the National Fish Hatchery System, and it enforces a broad array of federal laws relating to fish and wildlife conservation. It supports a range of wildlife and habitat conservation programs too numerous to describe here. This portfolio makes the FWS a very different agency from the National Park Service and the Bureau of Land Management; the Forest Service has some similarities, but its mandate is still more focused and more tethered to land management than is the FWS's.

Because the FWS has such a wide-ranging mandate, its structure differs from the other land management agencies (see Figure 6.5). The agency is divided both geographically and programmatically. The boundaries of the eight geographic regions mostly align with state boundaries, like the other agencies, but each agency's regional boundaries are remarkably different, reflecting not only land distribution but agency priorities as well. Program officers are distributed among the relevant regional offices but also report up to the national headquarters through a programmatic hierarchy. FWS law enforcement officers, for example, police the wildlife refuges according to agency policies prescribed at the national level, but these officers are distinct from land management staff, who are organized by region. Programmatic personnel are not evenly distributed across the various regions but are assigned according to regional needs.

One FWS organizational expedient not found in the other public land systems is the "complex." A refuge complex is an administrative grouping of multiple system parcels which share ecological features or management needs (for an example, see Figure 6.6). These properties are then managed from a single location and overseen by a project leader or complex manager and a single team of supporting staff members. One consequence of this approach is that some refuge properties may be unstaffed at any given moment, a fact which reflects both the FWS's relative lack of funding and the public's limited engagement with certain refuge parcels.[20]

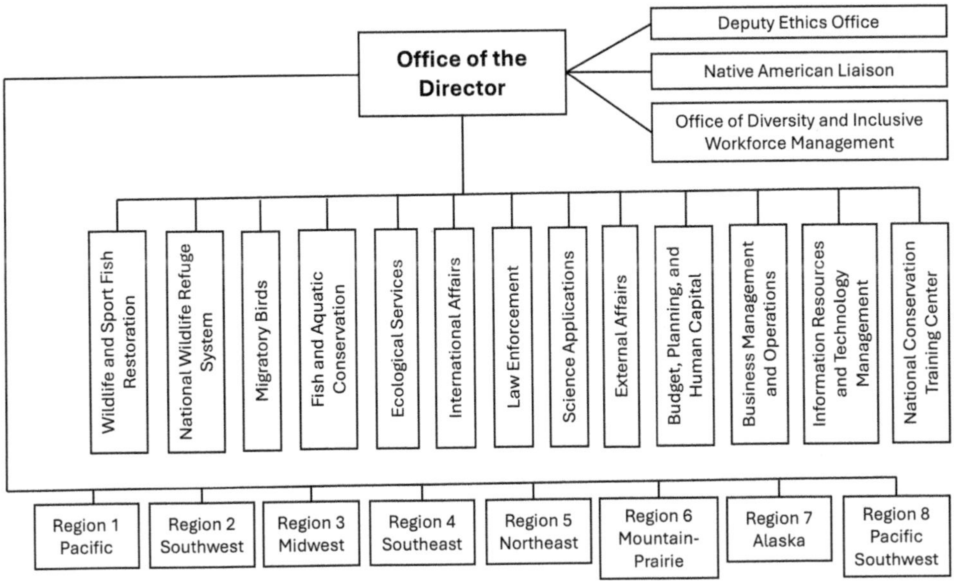

Figure 6.5 Organizational structure of the U.S. Fish and Wildlife Service. Source: Congressional Research Service.

Like the other federal land management agencies, the FWS must engage in a planning process for each refuge (or refuge complex). The process is spelled out in the Improvement Act and results in a Comprehensive Conservation Plan (CCP), which, in theory, guides the Service's decision making for a 15-year period, at which time the plan must be updated. In developing the plan, the team must consult not only with the general public but also with adjoining federal, state, local, and private landowners and affected state conservation agencies.[21] The Improvement Act specifically requires FWS planners to identify and describe:

- The purposes of each refuge comprising the planning unit;
- the distribution, migration patterns, and abundance of fish, wildlife, and plant populations and related habitats within the planning unit;
- the archaeological and cultural values of the planning unit;
- such areas within the planning unit that are suitable for use as administrative sites or visitor facilities;
- significant problems that may adversely affect the populations and habitats of fish, wildlife, and plants within the planning unit and the actions necessary to correct or mitigate such problems; and
- opportunities for compatible wildlife-dependent recreational uses.

Given these factors, CCPs can become very long documents, and producing them requires years of hard work from multi-member planning teams. Planning documents become the focal point of contested refuge management decisions. In certain cases, plans result in lawsuits against the FWS as affected parties allege that the agency disobeyed the law in one respect or another.

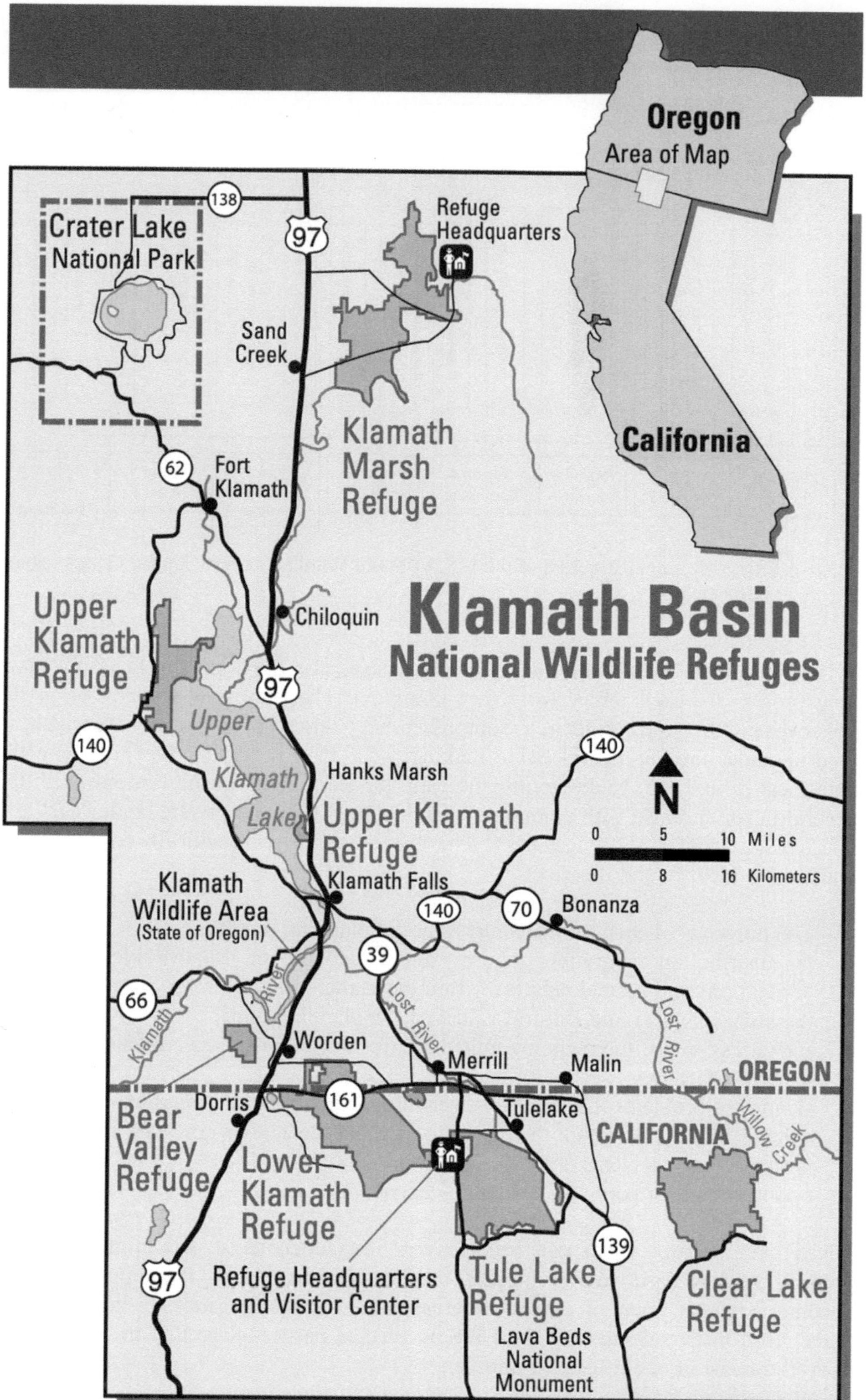

Figure 6.6 The Klamath Basin NWR Complex is made up of six individual refuges. Source: Fish and Wildlife Service.

What sorts of disputes might arise? Sometimes they have to do with water management, such as in the Klamath Basin NWR Complex, where agricultural irrigation often results in dry lakebeds within refuge boundaries—not ideal from a wetlands or aviary conservation perspective! Sometimes they have to do with oil and gas drilling, or perhaps the pace of cleanup of abandoned oil and gas infrastructure from drilling operations conducted long ago. One persistent source of conflict has to do with consumptive uses of wildlife refuges, such as the production of timber, hay, and grass. Under federal law, counties receive 25% of any federal revenue received for such commodities.[22] This tends to create a powerful local constituency in support of production and sale. The planning process becomes a forum for all of these issues and more; interested parties rightly understand planning documents to shape and constrain agency decisions while the plan is in effect. At the same time, those documents seldom bind agencies to concrete decisions. In most instances, the FWS retains the freedom to alter its course of action even after a plan has been formally adopted.

Conclusion

Alongside the other public lands systems—the National Forest System, the National Park System, and the System of Public Lands managed by the Bureau of Land Management—the National Wildlife Refuge System stands out. Outside of Alaska, wildlife refuges tend to be smaller and less impressively scenic than the Forests and the Parks, and more singularly focused on wildlife conservation (especially birds) than any of the other systems. Refuges are often popular with hunters, and the system has evolved to accommodate even non-wildlife-dependent uses, some of which conflict with the system's overarching purposes. But by and large, the refuge system remains an important and still-growing part of the public lands mosaic. As climate change and other environmental changes endanger wildlife populations and ecosystem health, the pressure on the FWS to preserve wildlife habitat will no doubt become more intense, but the agency's conservation expertise positions it well to respond to such challenges.

Study Questions

1. How does American ambivalence between purists and hunter-conservationists complicate the FWS's ability to maintain a coherent definition of "refuge?"
2. Evaluate the effectiveness of the Improvement Act's compatibility standard in balancing wildlife habitat protection with secondary uses, such as recreation.
3. What obstacles prevent the FWS from making a full transition toward a modern ecosystem management model?

Resources for Further Study

Fischman, Robert L. *The National Wildlife Refuges: Coordinating a Conservation System Through Law* (Washington, D.C.: Island Press, 2003).

Palomaki, Ashley. "The Battle Over Competing Land Uses Within National Wildlife Refuges: The Klamath River Basin as a Case Study." *Hastings West-Northwest Journal of Environmental Law & Policy* 20 (2014): 159–207.

Skillen, James R. *Federal Ecosystem Management: Its Rise, Fall, and Afterlife* (Lawrence, KS: University Press of Kansas, 2015).

U.S. Department of the Interior. *Flyways: Pioneering Waterfowl Management in North America* (Washington, D.C.: Government Printing Office, 1984).

Wilson, Robert M. *Seeking Refuge: Birds and Landscapes of the Pacific Flyway* (Seattle, WA: University of Washington Press, 2010).

Notes

1 Robert L. Fischman, *The National Wildlife Refuges: Coordinating a Conservation System Through Law* (Washington, D.C.: Island Press, 2003), 34.
2 16 U.S.C. § 718d(b).
3 16 U.S.C. § 668dd.
4 "Why Hunting Is Allowed on National Wildlife Refuges," available at https://www.fws.gov/story/why-hunting-allowed-refuges (last accessed June 18, 2025).
5 See generally James R. Skillen, *Federal Ecosystem Management: Its Rise, Fall, and Afterlife* (Lawrence, KS: University Press of Kansas, 2015). See also Fischman, *The National Wildlife Refuges*, 82–84.
6 See Fischman, *The National Wildlife Refuges*, 84.
7 16 U.S.C. § 715d.
8 16 U.S.C. § 668dd(a)(4)(D).
9 16 U.S.C. § 666f.
10 16 U.S.C. § 695l.
11 601 FW 1.15, citing 16 U.S.C. § 696. The U.S. Fish and Wildlife Service Manual (conventionally cited as "FW") can be found at https://www.fws.gov/policy-library/manuals. In the preceding citation, 601 refers to the "Part" number, and 1.15 to the specific provision within that part.
12 U.S. General Accounting Office, "National Wildlife Refuges: Continuing Problems with Incompatible Uses Call for Bold Action," Sept. 1989, GAO/RCED-89–196, 3.
13 See Fischman, *The National Wildlife Refuges*, 64–76. See also Box 4.2, infra.
14 16 U.S.C. § 668ee(4).
15 16 U.S.C. § 668dd(a)(3)(D).
16 603 FW 2.6.
17 *Wyoming v. United States*, 279 F.[3]d 1214, 1218 (10th Cir. 2002).
18 Later, it went farther, expanding the ban to all Alaskan wildlife refuges and extending it to cover *any* state-authorized "predator control" within the federal refuges. When Republicans took charge of Congress in 2017, however, the new Congress voted to abrogate the Refuges Rule under the Congressional Review Act, which permits Congress to nullify recently-enacted agency regulations.
19 *Safari Club Int'l v. Haaland*, 31 F.4th 1157, 1166 (9th Cir. 2022).
20 On funding, see Fischman, *The National Wildlife Refuges*, 33.
21 16 U.S.C. § 668dd(e)(3)(a).
22 16 U.S.C. § 715s.

7 The United States Forest Service

Introduction

This chapter introduces the National Forest System and its administration by the U.S. Forest Service. The National Forest System is quite unlike the park and refuge systems examined in the two previous chapters. While those systems are each dedicated to a dominant purpose, the Forest System is explicitly tasked with balancing multiple uses. Historically, the Service placed an unmistakable emphasis on timber production. But in recent years, recreational usage of the National Forests has increased dramatically. So too has ecological concern—and also wildfire. Today's Forest Service thus faces the extraordinary challenge of containing fires while preserving wildlife habitat; accommodating campers, hikers, and ranchers; and maintaining commercial relationships with timber harvesters and ski resorts.

In the pages that follow, we'll begin with the land itself—the nearly 200 million acres of national forests, which include much land that is *not* actually forested. We will discuss the major administrative distinctions within the system and from there will turn to the Forest Service's mission, the laws that guide it, and the agency's structure and organization.

The Lands of the National Forest System

Compare the National Forest System map (Figure 7.1) with the ecoregions map in Chapter One (Figure 1.2). In the eastern half of the United States, small National Forests dot the eastern temperate forest with no apparent pattern among the ecological regions. But in the western half of the country, patterns are evident. Large forest units dominate the Northwestern Forested Mountains, the Marine West Coast Forest, and the Temperate Sierra. There is also a swath of national forest along the southern edge of Mediterranean California.

The ecological variation within the National Forest System is striking. Consider the System's climatic diversity, for example. Visitors to the Tongass National Forests in southeastern Alaska should bring their best rain gear. Baranof Island, at the edge of the Gulf of Alaska, receives a stunning 235 inches of precipitation each year, and temperatures are moderate year-round, stabilized by proximity to the Pacific Ocean. In contrast, desert communities in the Coronado National Forest in Arizona receive as little as 10 inches of precipitation, and temperatures range more widely from 40°F to well over 100°F.

Even though millions of Americans visit the National Forest System, individual national forests do not enjoy the same name recognition as national parks like

DOI: 10.4324/9781003392514-7

Figure 7.1 The National Forest System. Map by Addison Arendt.

Yosemite, Yellowstone, Acadia, and Death Valley. There are several reasons for this, including the fact that national forests were neither created nor advertised primarily as recreational spaces. But perhaps another reason is that much of the public advertisement for recreation in national forests is by commercial enterprises operating under Forest Service lease. If you ski or snowboard, you have probably heard of resorts like Breckenridge, Copper Mountain, and Vail in Colorado; Mammoth in California; Timberline in Oregon; Mt. Baker in Washington; Loon Mountain in New Hampshire; or Sugarbush in Vermont. These are just a few of the 122 ski areas on national forest land, which, according to the Forest Service, provide more than half of all capacity for downhill skiing and snowboarding in the country.

National Forests

The National Forest System grew out of a profound fear that the United States would face a timber famine and degraded watersheds because of the rate at which private timber companies were deforesting the nation. As discussed in Chapter Three, the Forest Reserve Act of 1891 gave the president authority to set aside forest reserves to protect valuable timber resources and water quality. At that time, the federal government owned only limited acreage in the eastern half of the country, so presidents looked to the western federal estate, where they exercised this power with abandon. In 1905, the reserves were renamed "national forests," and by 1907, there were 159 national forests totaling 137 million acres (about 70% of the current National Forest System). Afraid of presidential excess, Congress revoked the president's authority to create national forests in 1907. The system continued to grow but through the much slower legislative process. Today there are a total of 154 national forests comprising 193 million acres across 40 states. The ten states without national forests are mostly on the East Coast (Connecticut, Delaware, Maryland, Massachusetts, New Jersey, and Rhode Island), along with several in the Great Plains (Iowa, Kansas, and North Dakota) and Hawaii.

Eastern National Forests and the Weeks Act of 1911

After terminating the president's unilateral authority to designate national forests, Congress in 1911 created a new mechanism for establishing national forests in the eastern half of the country. The context in the East was markedly different from that of the West: the northeastern forests and the Great Lakes forests had been almost entirely logged, and logging companies were quickly liquidating southern forests. Cut-and-run logging practices were often followed by intense fires, halting forest regeneration and leading to intensive soil erosion and devastating floods. These adverse effects prompted Congress to act. The Weeks Act of 1911 authorized the Secretary of Agriculture to purchase "forested, cut-over, or denuded lands within the watersheds of navigable streams as in his judgment may be necessary to the regulation of the flow of navigable streams and for the production of timber."[1] Whereas the western forests had been created as an act of protection, the Weeks Act authorized land purchases in the east for forest protection *or* forest restoration.

The process for creating Weeks Act forests varied greatly from the one used to designate national forests on public domain land. A National Forest Reservation Commission would identify areas of interest and draw gross boundaries around "forest purchase units." In most cases, the federal government obtained explicit permission to purchase land from state legislatures, and a national forest would only be created once the federal government had acquired enough land in the purchase area to allow for meaningful forest management. This could take considerable time. The Monongahela National Forest was established in 1920, some 9 years after creation of the purchase unit. Once a forest was established, the U.S. Forest Service worked to protect intact forest, extinguish fires, and reforest denuded areas. The Civilian Conservation Corps, a well-known New Deal program, played a critical role in this work during the 1930s.

Today, there are 51 national forests in the eastern half of the country, totaling about 25 million acres. Their history explains some of their distinctive characteristics. First, eastern national forests are generally much smaller than their western counterparts. Second, land ownership is fragmented in eastern national forests; in some cases, the federal government owns less than half of the land within the forests' gross boundaries. Third, when acquiring land, the federal government generally did not acquire the subsurface mineral estate, so states or private parties retained any valuable resources underneath these forests (which in later years would translate into disputes over coal, oil, and gas production within some national forests). Fourth, large-scale timber production wasn't immediately possible on many eastern forests, and the value of timber in these forests tends to be relatively low. This, combined with proximity to population centers in the East, meant that outdoor recreation became a significant priority in the first half of the twentieth century, long before similar pressures arose in the West.[2]

National Forests in Alaska

As with other federal land systems, the national forests in Alaska have unique legal, political, and economic dynamics compared to their counterparts in the Lower 48 states. Alaska has only two national forests, but they constitute more than 10% of the entire national forest system. They include most of the state's temperate rainforest, with dense stands of spruce, hemlock, and cedar (Marine West Coast Forest on the ecoregion map). President Roosevelt established the first federal forest reserve in Alaska in 1902, more than half a century before Alaskan statehood. In 1907, just before Congress revoked his authority to designate national forests, Roosevelt created the Tongass National Forest and the Chugach National Forest. At almost 17 million acres, today the Tongass dwarfs all other national forests and covers most of southeastern Alaska. It is fragmented by fjords and other waterways, which create some 11,000 miles of shoreline. Only a tiny fraction of the forest is accessible by road. This, combined with low population density in Alaska, has limited development in the forest over the last century.

The Tongass was established before significant timber demand existed and before Congress provided a basic framework for resolving federal, state, and Alaskan Native land claims. Boosters argued that the vast timber resources could build a significant lumber and paper industry, but that proved challenging. In 1947, Congress passed the Tongass Timber Act of 1947, which authorized the Forest Service to offer long-term (50-year) contracts with private companies, leading to several paper pulp mills. Timber harvest levels climbed quickly from less than 100 million board feet in 1947 to a high of almost 600 million in 1973. In 1971, Congress passed the Alaska Native Claims Settlement Act, which authorized the transfer of 45.5 million acres of land to Alaska Native corporations, including 550,000 acres of land in the Tongass National Forest. It also pledged to support the timber industry with continued federal timber harvesting.

But starting in 1980, Congress moved in a different direction. In the Alaska National Interest Lands Conservation Act (ANILCA) of 1980, Congress established fourteen wilderness areas, totaling 5.4 million acres, in which logging is prohibited. And in the Tongass Timber Reform Act of 1990, Congress restricted logging in another one million acres, while also removing earlier language that set a required level of timber harvesting. Today, the Tongass National Forest remains highly contentious, with presidential administrations alternatively working to expand or reduce the area open to logging.

National Grasslands

The U.S. Forest Service manages millions of acres of forested land. Despite the name, however, only about three quarters of the National Forest System's 193 million acres is actually forested. The remaining quarter includes alpine meadows high above tree line, semi-arid rangeland in the west, and prairie grasslands in the Great Plains. Yes, the *Forest* Service manages National *Grass*lands—twenty National Grasslands, totaling almost 4 million acres.

Like the Weeks Act forests, these grasslands were purchased by the federal government in response to dramatic land degradation. Most of the acres were bought between 1933 and 1946 to retire marginal agricultural land from crop production and restore it as productive rangeland for livestock, the goal being to return this acreage to private or state ownership once it had been stabilized. The Soil Conservation Service administered the land from 1938 until 1953, investing in extensive soil conservation projects. In 1953, the land was transferred to the Forest Service for temporary management, and in 1958, the federal government transferred some 6 million acres to state and private ownership. This left the Forest Service with about 3.8 million acres of grasslands, which were reclassified as "National Grasslands" in 1960.

For the agency, adding grasslands to the National *Forest* System raised significant management questions. On one hand, the Forest Service had managed livestock grazing on the national forests for a half century and was prepared to oversee grazing on the grasslands. But on the other hand, the grasslands did not fit the Forest Service's prevailing emphasis on timber production. As one forester explained,

> Many Forest Service administrators interpreted this [reclassification] to mean to manage the Grasslands the same as National Forests. Others were of the opinion that the National Grasslands were only big cattle pastures (existing) solely for the use and benefit of the grazing users. Still others saw the grasslands as a bastard child and the best alternative was to dispose of them.[3]

With the exception of one additional grassland that Congress created in 1991—the Butte Valley National Grassland in northern California—the national grasslands are located in the Great Plains ecoregion, filling a geographic gap between eastern and western national forests. They are managed according to the same multiple-use mandate as the national forests, producing wildlife habitat, forage for domestic livestock, oil and gas, and recreational opportunities.

Research and Experimental Areas

Almost from its inception, the Forest Service has been engaged in applied research to understand forest and grassland dynamics and to improve various aspects of natural resource development. From the first experimental forest in 1908 (Fort Valley Experimental Forest in Arizona) to the most recent addition in 2007 (Sagehen Experimental Forest in California), the network of experimental areas has grown to encompass most forest types in the country as well as numerous grassland types. Today there are 77 experimental areas covering about 485,000 acres, and researchers use additional research areas within the National Forest System to further their work.

Experimental forests and rangelands have had a significant impact on both research and management in the United States. Researchers have tested forestry practices, watershed management techniques, decomposition and carbon sequestration, forest product development, and much more. Research in these areas has helped reframe the dominant paradigm of forest management. The H.J. Andrews Experimental Forest in Oregon, for example, was established in 1948 to study the best methods for clear-cutting old growth forests and replacing them with young, faster-growing forests. As researchers shifted their attention to long-range ecosystem dynamics, they helped lead the Forest Service's adoption of a more comprehensive approach called ecosystem management in the 1990s.[4]

Other Designations

The National Forest System includes a range of other designations (see Table 7.1), which whether legislative or administrative in origin generally narrow the managerial discretion the agency ordinarily enjoys. For example, Congress has created 449 wilderness areas on Forest Service lands, totaling 18% of the entire National Forest System. Wilderness, by definition, must remain free of permanent improvements such as roads, and the use of mechanized equipment is prohibited as well. This places significant limits on the types of commercial resource development and recreation that the Forest Service can authorize. In addition, the Forest Service has added an administrative designation of "roadless" to almost 60 million acres of land, which also limits authorized uses. These administrative designations can be highly contentious and dynamic because they restrict commercial use.

Roads

It might seem odd to have a section on roads in a discussion of the National Forest System. But roads deserve special mention, not least because the Forest Service estimates that there are 370,000 to 380,000 miles of roads in the system. Compare this to the National Wildlife Refuge System's 4,800 miles of roads and the National Park System's 13,100 miles.

Table 7.1 Additional Designations in the National Forest System.

Name	*Number*	*Extent of Coverage*
Wilderness Areas	449	36,669,333 acres
National Wild and Scenic Rivers	134	~5,000 miles
National Recreation Areas	23	3,246,330 acres
National Game Refuges or Wildlife Preserves	22	1,217,680 acres
National Monuments	14	4,667,352 acres

Not all roads are created equally. There are state highways running through National Forests, and the Forest Service manages an extensive network of paved roads. But most of this vast road system is unpaved and receives little to no maintenance. Indeed, the agency estimates that less than 20% of its road miles meet environmental and safety standards. The problem is that most roads were built by the agency or private companies for timber harvesting, and they were not engineered as permanent, public roads. Combine low construction standards with a lack of maintenance and you get a road system that is literally crumbling, creating immense problems of erosion and public safety. The agency estimates that it has a road maintenance backlog of about $8.4 billion, which fuels public pressure to block new road construction.[5] The Great American Outdoors Act of 2020 provided supplemental funding for road maintenance, but not enough to bring the entire system up to standards.

The Forest Service addresses road construction and maintenance in each forest's or grassland's land management plan. One of the contentious questions in many land management plans is whether to close roads to public use or even decommission roads in order to restore the land along its path.

The Mission of the U.S. Forest Service

The Forest Service has a multiple-use, sustained-yield mission that gives it broad discretion in balancing public demands on the Nation Forest System.[6] As we'll explain later in this chapter, the clearest articulation of this mission is found in the Multiple-Use Sustained Yield Act of 1960, in which Congress declared, "the national forests are established and shall be administered for outdoor recreation, range, timber, watershed, and wildlife and fish purposes."[7] But even before 1960, the Forest Service's mission was understood to be broad enough to encompass divergent and competing views. In 1905 the Service's first chief, Gifford Pinchot, famously charged the agency with providing "the greatest good to the greatest number for the longest time." This mandate, however, is particularly challenging when Americans have different ideas about what constitutes "the greatest good" and which "greatest number" should benefit.

The difficulty here isn't just that some people want to hunt and some people want to cut timber, though that is a challenge in and of itself. Over the last 120 years, the way Americans *see* forests and grasslands has expanded and splintered. Gifford Pinchot was a leading environmental thinker in his day and deeply committed to protecting the health of the national forests. He was an avid angler who loved the outdoors. But based on the science of the time and his European training in forestry, he saw the national forests in agricultural terms of production. In his 1910 book, *Breaking New Ground*, he explained it this way:

> Forestry is Tree Farming. Forestry is handling trees so that one crop follows another. To grow trees as a crop is Forestry. Trees may be grown as a crop just as corn may be grown as a crop. The farmer gets crop after crop of corn, oats, wheat, cotton, tobacco, and hay from his farm. The forester gets crop after crop of logs, cordwood, shingles, poles, or railroad ties from his forest, and even some return from regulated grazing … . Farmer and forester alike get a lot of other products on the side. Good farming yields also such things as butter, eggs, apples, calves. Good Forestry, in addition to lumber, firewood, and other produce, yields such services as regulation of stream flow, protection against erosion, and some influence on

> climate … . The purpose of Forestry, then, is to make the forest produce the largest possible amount of whatever crop or service will be most useful, and keep on producing it for generation after generation of men and trees.[8]

Pinchot's production framework is still widely held, even if some of its advocates have added other "products" to his list, such as fish and wildlife habitat, recreational opportunities, and even carbon sequestration.

But there are other lenses through which to view forests and grasslands, including an ecological lens that was not yet available to Pinchot when he became the first Forest Service chief. The most influential advocate of this newer ecological perspective was a forester hired by Pinchot's agency in 1909, Aldo Leopold. He had gone to Pinchot's school of forestry at Yale and initially embraced Pinchot's utilitarian orientation. But as his career unfolded, he increasingly found the Forest Service's understanding of the natural world to be dangerously reductive. Leopold began to advocate for a more holistic view of forests and grasslands, arguing that the goal of land management was to preserve land health, which he defined as its capacity for self-renewal. This, he insisted, required seeing forests and grasslands as ecological communities rather than production facilities:

> In the field [of conservation] one group (A) regards the land as soil, and its function as commodity-production; another group (B) regards the land as a biota, and its function as something broader … . In my own field, forestry, group A is quite content to grow trees like cabbages, with cellulose as the basic forest commodity. It feels no inhibition against violence; its ideology is agronomic. Group B, on the other hand, sees forestry as fundamentally different from agronomy because it employs natural species, and manages a natural environment rather than creating an artificial one … . It worries on biotic as well as economic grounds about the loss of species.[9]

This led to Leopold's famous ethical axiom: "A thing is right when it tends to preserve the integrity, stability, and beauty of the biotic community. It is wrong when it tends otherwise."[10]

The tension between these two perspectives has been magnified by developments in ecology and related sciences, which have provided far more support for Leopold's view than Pinchot's. For example, fire ecology has demonstrated fire's vital ecological role in many forest and grassland systems, challenging Pinchot's view that fire was only a destructive process that wasted valuable timber. Wildlife biology has demonstrated the critical ecological role that predators play in sustaining healthy game populations, challenging the notion that they are simply competition for hunters.

To be clear, the two views so briefly outlined here—forests as tree farms, forests as biotic communities—are not binary. Leopold's view would not require the Forest Service to halt economic resource development or turn the entire National Forest System into federally designated wilderness; Leopold himself was an avid hunter and angler who supported timber production and other economic land uses. There is a spectrum of views that are held by both Forest Service employees and users of the National Forest System, and the Forest Service's authorizing legislation provides support for this full spectrum. This range of motion sounds nice on paper, but in practice it means that the agency's multiple-use mission is never completely settled. Opposing forces can always contend for a different managerial focus or policy.

This missional flexibility means that Forest Service management evolves over time in response to public pressure and legislative change, as well as extensive litigation. One of the most striking shifts has been the decline in timber production and the rise of recreation. In the 1960s, 1970s, and 1980s, the Forest Service sold about 8 to 12 *billion* board feet of timber annually from the national forests. But public pressure and a variety of federal statutes led to a steep decline in timber production. Over the last 20 years, national forest timber production has hovered around 2 to 3 billion board feet annually. Meanwhile, recreational use of the national forests has exploded, more than tripling between 1960 and the Forest Service's 100th anniversary in 2005. These shifts reflect and drive significant economic and cultural changes, particularly in the eleven western states, and help explain why national forest management is often contentious.

In sum, today's Forest Service supports economic development of natural resources (logging, grazing, etc.), a broad range of recreational activities (camping, hiking, hunting, fishing, etc.), and biodiversity protection. The exact balance is determined in some cases by other congressional legislation, such as wilderness designations, but otherwise it is left to the agency's discretion and determined through a detailed land use planning process.

National Forest System Law

Unlike the National Park Service, which has a single organic act written in 1916, the Forest Service's overarching mandate is more like an edited collection of laws written in 1897, 1960, and 1976, with additional supporting legislation in between. The best way to understand the agency's legal context, then, is to look at how its mandate expanded through these three pieces of legislation.

The Organic Act of 1897

The Organic Act of 1897 served the agency well in its early decades, when use of most national forests was relatively low. Significant parts of this law have been repealed over time, but the U.S. Code still contains its articulation of the primary purposes of the national forests:

> to improve and protect the forests within the boundaries, or for the purpose of securing favorable conditions of water flows, and to further a continuous supply of timber for the use and necessities of citizens of the United States.[11]

In its historical context, this law was directing the Forest Service to protect watersheds from soil erosion and flooding and to produce a sustained yield of timber. There were, in other words, just two primary goals in the Forest Service's mission at the turn of the twentieth century. The Act gave and continues to give the Forest Service latitude in determining the types of uses compatible with these two goals, from domestic livestock grazing to outdoor recreation.

Multiple-Use Sustained Yield Act of 1960

As use of the National Forest System increased, thanks to expansive timber harvesting and an explosion in outdoor recreation, the agency quickly outgrew its original Organic Act. Indeed, the 1897 Act had said nothing about recreation, livestock grazing, or fish and wildlife management, even though the Forest Service was confident that it could ably serve the full array of public demands if it was given the authority to do so.

In 1960, Congress added substantially to the Forest Service's missional authority in the Multiple-Use Sustained Yield Act (MUSY), in many ways simply catching up with the agency's management realities. The new policy in MUSY was to manage the National Forest System "for outdoor recreation, range, timber, watershed, and wildlife and fish purposes."[12] Lest it be misunderstood as mandating a hierarchy of uses, Congress put the uses in alphabetical order—although some accused Congress of signaling priorities by adding the word "outdoor" to "recreation" and by reversing the customary "fish and wildlife" to "wildlife and fish."

The terms "multiple use" and "sustained yield" are critical. Multiple use means

> the management of all the various renewable surface resources of the national forests so that they are utilized in the combination that will best meet the needs of the American people; making the most judicious use of the land for some or all of these resources or related services over areas large enough to provide sufficient latitude for periodic adjustments in use to conform to changing needs and conditions.[13]

Sustained yield is defined to mean "the achievement and maintenance in perpetuity of a high-level annual or regular periodic output of the various renewable resources of the national forests without impairment of the productivity of the land."[14] Again, both definitions avoid any clear hierarchy of management outcomes.

But despite Congress's attempt at neutrality, legislators still found ways to insist on particular priorities. Mostly this occurred by way of the annual appropriations process. In the 1960s, those appropriations made it clear that timber production remained at the top of the list, even though Congress increased funding for campgrounds and other visitor services. The Forest Service's staff was still dominated by foresters, and public trust in the agency waned in the next decade, driven by growing criticism from the environmental movement.

BOX 7.1 Monongahela National Forest Clear-Cutting Controversy

In the post-war period the Forest Service faced enormous pressure to increase timber production, which pushed it even further down the road toward Pinchot's agronomic view. The agency wanted to increase harvesting efficiency and tree growth for future harvests. In 1964, it announced that it was largely shifting away from selective timber harvesting, where loggers cut only mature trees of a particular diameter, leaving younger trees in place. Instead, the Service intended to utilize clear-cutting, in which all trees are removed. This was a less expensive way for logging companies to harvest timber, and it allowed the Forest Service to replant preferred tree species with spacing that would produce maximum tree growth.

But when the agency began offering large clear-cut sales on the Monongahela National Forest in the mid- to late 1960s, it quickly became clear that it had underestimated public opposition. In 1973, the Izaak Walton League sued the Forest Service on behalf of hunters, whose favorite turkey-hunting spot had been clear-cut. Their legal argument centered on the Organic Act of 1897, which authorized the Secretary of Agriculture to sell "dead, matured, or large growth" trees and which required the Forest Service to mark each individual tree that was harvested. The government argued that MUSY "has by implication repealed the restrictive provisions of the Organic Act," but the trial court and the appeals court sided with the Izaak Walton League.[15] The 1975 decision put an immediate halt to the Forest Service's preferred method of timber harvesting. It became clear that the Service's legal authority needed revision: the industry argued that clear-cutting was necessary in some instances, while environmentalists demanded greater attention to environmental concerns. The resulting pressure induced Congress to pass the National Forest Management Act of 1976, which was so detailed and comprehensive that it was, in effect, a new organic act for the Forest Service.

National Forest Management Act of 1976

In 1976, in response to the Monongahela dispute (see Box 7.1) and burgeoning environmental pressures, Congress passed the National Forest Management Act (NFMA; pronounced "NIF-muh"). NFMA reaffirmed the Forest Service's over-arching multiple-use, sustained-yield mission in managing the National Forest System, directing the agency "to provide for recreation (including wilderness), range, timber, watershed, wildlife, and fish."[16] You'll note the amusing shuffling of the list from MUSY to NFMA, but again no hierarchy of uses is prescribed. But the similarities between MUSY and NFMA are deceptive; in fact, NFMA represented a major change in the law of the Forest Service. Its most important development was a detailed mandate for land use planning. NFMA directed the agency to prepare (and subsequently to update and revise) a comprehensive plan for each unit of the National Forest System. This process involved several requirements that have reshaped Forest Service management.

First, forest plans must rest on interdisciplinary analysis by a range of experts to ensure that the agency is considering both the economic and environmental implications of its proposed resource management. The word "interdisciplinary" sounds bland, but in this context it was anything but. Throughout the first half of the 1900s, the Forest Service had been allowed to make decisions solely on the advice of its professional foresters. NFMA threw open the doors to other disciplines and scientists. To meet NFMA's requirements, the Forest Service had to diversify its workforce and add a host of other specialists.

Second, the plans must include robust public participation. The agency must publicize its planning process, hold public meetings in or near the planning area, publish a draft plan, and respond to public feedback. The public participation requirements ensure that every interested American has access to the Forest Service decision-making process, at least at the level of land management plans. This has altered the mix of public demands that the Forest Service hears in its planning process. An environmental studies professor in Michigan and a law professor at Notre Dame, for example, can participate quite easily in the development of a land use plan for the Flathead National Forest in Montana.

Third, NFMA added new environmental protection mandates. These included specific protection mandates for timber harvesting and a general mandate to "provide for diversity of plant and animal communities based on the suitability and capability of the specific land area in order to meet overall multiple-use objectives."[17] The Forest Service has interpreted this to mean that its plans must "provide the ecological conditions to both maintain the diversity of plant and animal communities and support the persistence of most native species in the plan area."[18] Note that the diversity requirement implicates the management of *all* species, not just the endangered or threatened species protected by the Endangered Species Act. There is little doubt that NFMA's diversity mandate has contributed to a reduction in timber harvesting in the national forests.

All three of these analytical requirements dovetail nicely with the National Environmental Policy Act of 1969 (covered in Chapter Four), which was still relatively new at the time NFMA was enacted. The interdisciplinary and diversity analyses required by NFMA can be folded directly into the NEPA process, which in turn requires and facilitates public participation. This is why NEPA documents, as noted previously, are sometimes inseparable from other planning or analytical documents.

Additional Land Designations

Congressional legislation and presidential proclamations have placed additional designations on some lands within the National Forest System. Of these, the most significant are the agency's 449 wilderness areas, totaling over 36 million acres. The Forest Service manages more than half of all the wilderness units in the entire United States, representing a full third of the system's acreage. These areas are still multiple-use areas in the sense that they provide recreation, watershed protection, and fish and wildlife habitat (and even livestock grazing in some instances). But as Chapter Four explained, the Wilderness Act places significant limitations on other uses, prohibiting roads and any activity that requires mechanized equipment.

Other examples of special designations are National Recreation Areas (which are created by Congress) and National Monuments (which may be created by Congress or by a president). These designations require the Forest Service to depart from its ordinary multiple-use mandate and instead to manage land for a dominant use, whether it is recreation or the preservation of historic and scientific interest. Needless to say, this leads to very different managerial priorities and places obvious limits on activities like logging or mining.

Forest Service Regulations

Like many agencies, from time to time the Forest Service must issue its own regulations—either because a congressional statute requires it, or because the agency wishes to establish a consistent policy for implementing a statute. (If you're curious, Forest Service regulations are compiled in Chapter 2 of Title 36 of the Code of Federal Regulations.) The rulemaking process can be highly contentious. One recurring challenge the Service faces has to do with the regulations by which it implements NFMA's planning requirements. The "Planning Rule," as it's called, is important because it establishes rules that will apply across all units of the National Forest System. Changes to the Planning Rule have been a regular source of both political tension and litigation.

The Forest Service published its first set of planning regulations in 1979 and revised them in 1982. Subsequent revisions to the planning regulations have been fraught. Just before leaving office, the Clinton Administration finalized a new Planning Rule, leading to lawsuits from both environmental and industry groups. The George W. Bush Administration postponed that 2000 rule and adopted its own regulations in 2005. But a judge blocked the Bush regulations, ruling that the Forest Service had violated procedural requirements in both NFMA and NEPA. After additional work, the Forest Service adopted very similar regulations in 2008, only to have those enjoined for violating NEPA and the ESA. Each time the regulations were struck down, the agency reverted to its 1982 regulations. The agency at last succeeded in updating regulations in 2012, and that 2012 Planning Rule remains operative—although later administrations have continued to consider even more changes.

What is so contentious about the Planning Rule, given that it primarily governs the planning *process* rather than dictating outcomes? It is contentious because procedural changes affect the information that the agency receives, the priorities it sets, and the ways that interested parties may participate in the agency's decision making. These considerations can influence plan outcomes. Among its other features, the 2012 Planning Rule emphasizes multidisciplinary science, the protection of forests' ecological services, and the consideration of climate change. As such, the Rule is thought to elevate ecological protection in the national forests and grasslands—for now. Like any regulation (or any statute, for that matter!), the 2012 Planning Rule might be repealed or replaced by the time you read this.

Forest Plans

Congressional legislation, executive orders, and agency regulations all have to be translated to specific units of the National Forest System, which happens in the land management plans mentioned above. The Service seeks to support multiple uses, but clearly not every use is possible or appropriate on every acre of the System. In each land management plan, the agency explains the existing and desired ecological conditions, subdivides the plan area into smaller units, and sets objectives for each area. For example, the land management plan for the Helena-Lewis and Clark National Forest covers just under 2.9 million acres, subdivided into 10 distinct areas. The area contains a range of special designations, including wilderness, inventoried roadless areas, wild and scenic rivers, and research natural areas, each of which impacts management options.

Throughout the planning area, the Forest Service considers the possibility of timber production, differentiating between areas suitable for and areas not suitable for timber harvesting. In the case of the Helena-Lewis and Clark National Forest, the agency has determined that roughly 2.5 million acres are not suitable for timber harvesting for legal or technical reasons (e.g., wilderness designation or steep slopes) or because timber harvesting is incompatible with other management objectives (e.g., wildlife habitat). To be sure, the forest has timber production goals—but given the vast acreage not suitable for timber harvesting, those goals must be met from the remaining 368,563 acres. This type of analysis is done for every unit of the National Forest System.

The staff of each forest and grassland uses its land management plan as a long-term guide and then plans individual projects in accordance with the plan. But the agency can only carry out activities that Congress will fund, so specific management actions vary with the agency's annual appropriations. For example, one long-term trend is that Congress is providing more funding for hazardous fuels reduction and correspondingly less money for timber harvesting.

National Forest System Operations

As we explained at the beginning of this chapter, our focus is on federal public lands, so this chapter is limited largely to the Forest Service's management of the National Forest System, with only a brief mention of the agency's research stations. The agency as a whole employs roughly 30,000 permanent employees; it also hires approximately 10,000 additional seasonal staff.

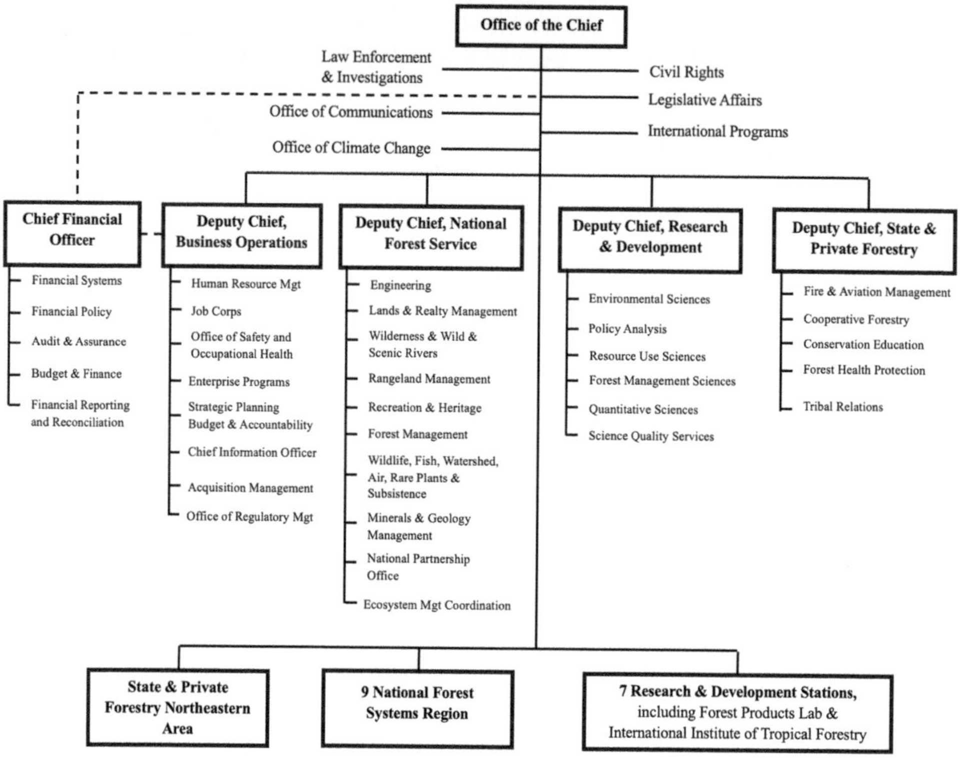

Figure 7.2 The organizational structure of the U.S. Forest Service. Source: Forest Service.

The Forest Service is led by a chief who is appointed by the Secretary of Agriculture. In contrast to the other three land agencies, the Forest Service chief is not a political appointee requiring Senate confirmation. The chiefs have almost all risen through the ranks of the agency, and three-quarters of the chiefs have educational backgrounds and career experience in forestry. This has somewhat insulated the Service from the shifting politics in the White House and the Senate. There are four deputy chiefs in the agency; they oversee business operations, the National Forest System, research and development, and state and private forestry. The majority of Forest Service employees work in the National Forest System.

The system is divided into nine regions, each overseen by a regional forester whose office, historically, was located within the region.[19] For reasons explained above, the Southern and Eastern regions are the smallest in terms of national forest and grassland acreage, with about 13 million and 12 million acres respectively. The largest is the Intermountain Region at about 31 million acres. The regional foresters direct the supervisors responsible for individual national forest or grassland units.

Staffing

As Forest Service management priorities have shifted, so too has the agency's employment profile. When timber production was the top priority, the agency was dominated

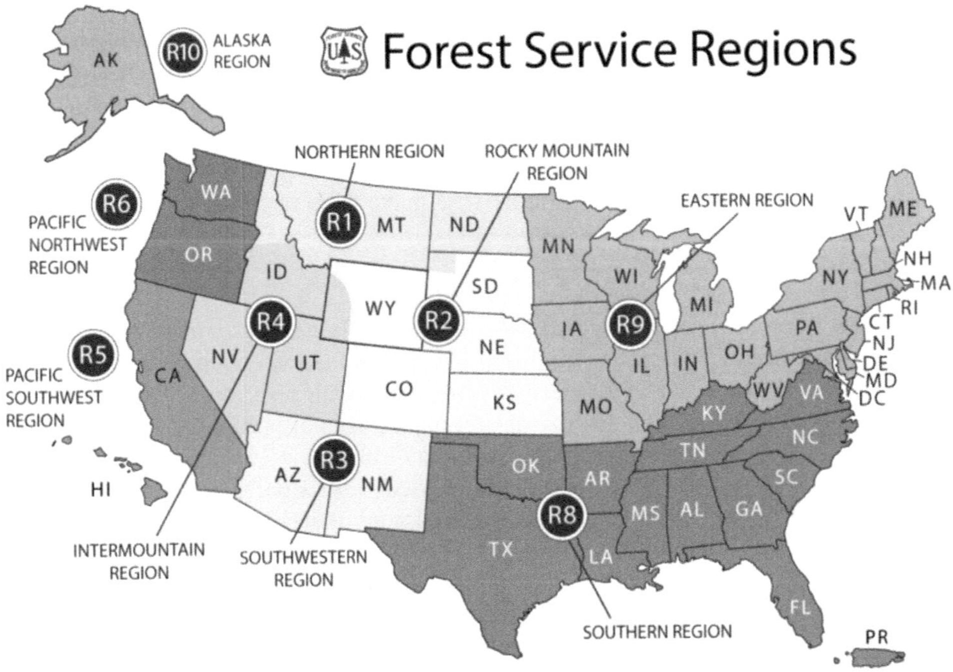

Figure 7.3 Regions of the Forest Service. Region 1: Northern; 2: Rocky Mountain; 3: Southwestern; 4: Intermountain; 5: Pacific Southwest; 6: Pacific Northwest; 7: merged into Eastern in 1965; 8: Southern; 9: Eastern; 10: Alaska. Source: Forest Service.

by foresters, forestry technicians, and road-building engineers. But timber production plummeted in the 1990s, and forestry-related positions followed. In the last 20 years, the Forest Service has faced a growing wildfire crisis, and fire staff has ballooned. The National Association of Forest Service Retirees found that between 1992 and 2018, forestry and engineering positions dropped by 50–75% (see Table 7.2). In that same time, the primary growth was in direct fire positions and botany positions, and many of the latter are related to fire management.

Table 7.2 Forest Service staff roles in 1992 and 2018: changes by number and percentage.

Skill	*1992*	*2018*	*Loss/Gain*	*Percentage*
Forester	5,143	1,337	-3,806	-74%
Engineering Technician	1,905	525	-1,380	-72%
Professional Engineer	1,566	761	-805	-51%
Forestry Technician	7,910	4,000	-3,910	-49%
Interdisciplinary Team Leader	7,604	4,175	-3,429	-45%
Wildlife Biologist	930	568	-362	-38%
Fisheries Biologist	371	272	-99	-27%
Botanist	124	159	+35	+28%
Fire	2,800	6,500	+3,700	+132%

As with the other federal land agencies, the Forest Service employs a wide range of seasonal workers each year, although the lion's share of these are related to fire. The Service has also increased the number of "permanent seasonal" positions, which guarantee 6 months of work each year and make the employee eligible for important benefits year-round. The agency also has direct hiring authorities that allow it to hire recent graduates, allowing young graduates to bypass competitive hiring processes in which their lack of experience would otherwise make them uncompetitive.

Budget

This brings us to the agency's budget, which is driven largely by annual, discretionary appropriations from Congress. As discussed in Chapter Four, Congress has been largely gridlocked for decades, unable to enact major public lands legislation despite repeated attempts by both parties. For this reason, recent changes in public lands management have instead come about through executive orders, rulemaking, or appropriations.

Overall, the Forest Service budget has increased substantially over the last several decades. The Congressional Research Service found that the Forest Service's total appropriations increased by 36% from 2011 to 2020 (from $6.0 billion to $8.2 billion, in 2020 dollars). But as we just saw, employment in forestry-related categories has fallen precipitously. So where did all of the additional funding go? The short answer is wildland fire management. In the twenty-first century, federal land agencies have struggled to adapt to changing fire regimes, and a growing percentage of their budgets has been consumed by fire suppression costs.[20]

Indeed, over the last two decades, wildland fire management has consumed roughly half of the agency's entire budget. This puts enormous strain on programs unrelated to fire. Even in years in which its other programs receive ample funding from Congress, the Forest Service has had to borrow funding from those programs to cover the mushrooming costs of fire management. In some years, Congress has later provided emergency supplemental funding, but the Service has often had to pause spending in everything from recreation to wildlife management while it waited for Congress to act.

In 2018, Congress provided a "fire fix" to address this budgetary problem. For 8 years, the Forest Service was given access to an additional $1.95 billion annually for wildland fire management. In 2021, Congress provided even more funding focused on reducing the risk of catastrophic wildfire. The Infrastructure Investment and Jobs Act of 2021 provided roughly $3 billion more for the Forest Service over 5 years. The additional funding was directed toward prescribed burning, mechanical fuels reduction, and other forest treatments, as well as higher pay and additional benefits for wildland firefighters. With these additions, Forest Service appropriations increased to $9.3 billion in the 2024 fiscal year, with a full 50% dedicated to fire-related programs.

There is one other budget issue that is critical to understanding the political pressures faced by the Forest Service. Although most of the agency's funding comes through annual appropriations bills, a small percentage of Forest Service appropriations are so-called "mandatory appropriations" or "mandatory spending." This funding authorizes specific programs over longer periods of time, freeing the agency from the need to seek annual approval from Congress. The Forest Service has about two dozen mandatory appropriations accounts that make up about a tenth of its total budget. Many of the mandatory accounts are revenue-sharing programs that benefit states and counties or support specific work in the national forests. For example,

25% of the gross revenue from national forests and grasslands (calculated using a 7-year average) is paid directly to counties that contain national forests and grasslands. These funds support roads and schools. Another example is the Knutson-Vandenberg Trust Fund, which allows the Forest Service to use some timber sale funds for reforestation and other forest management activities.

These mandatory accounts are politically important because of their fiscal impact on counties with federal public lands. Many counties in the West receive millions of dollars each year from the Forest Service, creating an obvious incentive for such counties to demand increased timber production. When timber production declines, as it did precipitously in the Pacific Northwest in the 1990s, many counties faced devastating budget shortfalls. This in turn has an enormous effect on the agency's relationship with county and state governments and with congressional delegations from affected states.[21]

Conclusion

The U.S. Forest Service is primarily a forest manager, stewarding vast swaths of mountainous forest in Alaska and the West and smaller but significant forest areas in the east. It also manages grasslands throughout the country. But what forest and grassland management *means* and what the agency prioritizes has changed significantly over time. The agency spent much of its history focusing on multiple human uses of the forests and grasslands, with a primary emphasis on economically valuable forest products and a secondary focus on recreational opportunities. But since the 1970s, it has expanded its management objectives significantly to include protection of ecosystems and the ecological benefits they provide. A growing percentage of the National Forest System is closed to logging. Sometimes the agency itself, goaded by its research staff, has pushed in this new direction; at other times, it has been forced in this direction by Congress or the courts. Meanwhile, the cost of fire management has exploded, threatening to crowd out other priorities. No one knows the future, but you can bet that the Forest Service's multiple-use mandate will continue to present countless challenges as the agency serves a population that wants dramatically different things from the National Forest System.

Study Questions

1 How does the operational reality of the U.S. Forest Service affect its legislatively-mandated mission to balance multiple uses?
2 Is the pursuit of sustained yield compatible with preserving the integrity, stability, and beauty of a natural ecosystem?
3 When litigation is required to coax the Forest Service to modify its priorities—such as the clear-cutting controversy leading to the National Forest Management Act (NFMA)—does this imply that the agency's discretionary authority is structurally flawed? Or is litigation just an inevitable political consequence of the multiple-use mandate?

Resources for Further Study

Leopold, Aldo. *A Sand County Almanac* (New York, NY: Oxford University Press, 1966).
Riddle, Anne A. *National Forest System Management: Overview and Issues for Congress* (Washington, D.C.: Congressional Research Service, 2023).

Thomas, Jack Ward. *The Journals of a Forest Service Chief* (Durham, N.C.: Forest History Society, 2004).

West, Terry. "USDA Forest Service Management of the National Grasslands." *Agricultural History* 64, no. 2 (1990): 86–98.

Williams, Gerald W. *The U.S. Forest Service in the Pacific Northwest: A History* (Corvallis, OR: Oregon State University Press, 2009).

Notes

1 Public Law No. 94–588, § 17(a)(3).

2 William E. Shands, "The Lands Nobody Wanted: The Legacy of Eastern National Forests," in *The Origins of the National Forests*, ed. Harold K. Steen (Durham, N.C.: Forest History Society, 1992).

3 Deputy Regional Forester Rupp, quoted in Terry West, "USDA Forest Service Management of the National Grasslands," *Agricultural History* 64, no. 9 (1990): 86–98, 89.

4 See James R. Skillen, *Federal Ecosystem Management: Its Rise, Fall, and Afterlife* (Lawrence, KS: University Press of Kansas, 2015), 142–149.

5 See the Forest Service's Road Management Website, located at https://www.fs.usda.gov/eng/road_mgt/overview.shtml.

6 While this chapter focuses on the agency's management of the National Forest System, about 15% of the agency's 30,000 permanent employees work outside the System, supporting state and private forestry and international development.

7 16 U.S.C. § 528.

8 Gifford Pinchot, *Breaking New Ground* (Washington, D.C.: Island Press, 1998), 31–32.

9 Aldo Leopold, *A Sand County Almanac* (New York, NY: Oxford University Press, 1966), 258–259.

10 Leopold, *A Sand County Almanac.*

11 16 U.S.C. § 475.

12 16 U.S.C. § 531(a).

13 16 U.S.C. § 528.

14 16 U.S.C. § 531(b).

15 The appellate court's decision is *West Virginia Division of Izaak Walton League of America v. Butz*, 522 F.2d 945 (4th Cir., 1975).

16 16 U.S.C. § 1604(e)(1).

17 16 U.S.C. § 1604(g)(3)(b).

18 36 C.F.R. § 219.9.

19 At the date of this book's publication, the Trump Administration has declared its intention to close the regional offices. If this comes about, presumably the entire agency will be managed from its headquarters in Washington, D.C.

20 The Forest Service has its own wildland fire management budget, while the other three land agencies receive consolidated appropriations through the Interior Department.

21 Those interested in these dynamics may wish to examine a project called "Follow the Money: A Spatial History of In-Lieu Programs for Western Federal Lands," which is part of the spatial history project at Stanford University. The project allows the viewer to observe payment fluctuations within specific programs.

8 The Bureau of Land Management

Introduction

The Bureau of Land Management (BLM) occupies a fascinating position. It manages more acreage than any other federal agency, yet few are familiar with its mission or purpose. Many easterners are unaware even of its existence, and the lands it stewards are decidedly less scenic and more arid than most other public lands. If you're passing through a nondescript stretch of Nevada or Wyoming, with no water or trees in sight, there's a good chance you're in BLM territory. But make no mistake: in the American West, the BLM is very important indeed. It manages vast mineral resources, including oil, gas, and coal reserves worth billions of dollars. It administers grazing rights that cover more land than any other single federal land use. This chapter will explain these apparent contradictions, first describing the land under BLM management before turning to the agency's mission and the laws that govern it. The chapter ends with a brief discussion of the agency's operations and organization.

The Land Managed by the BLM

The BLM's landholdings are quite unlike those of the Park Service, Forest Service, or Fish and Wildlife Service. BLM land includes vast swaths of desert and semi-arid land in eleven western states, including more than 60% of all surface land in Nevada and nearly 85 million acres of tundra and taiga in Alaska. In many places, BLM land is divided into scattered parcels, and only rarely are BLM tracts bestowed with proper names like national parks and national forests. Road maps and atlases of the western states tend to show national forests, monuments, parks, recreation areas, and wildlife refuges, but not BLM land. As you drive across the American West, you often cannot distinguish BLM land from land in state and private ownership. You are likely to see highway signs announcing that you have entered or exited a national forest, but very few BLM boundaries have any signage at all. The official, collective name of BLM lands is also rather bland—the "National System of Public Lands"—and can be easily misinterpreted as a reference to *all* federal public lands.

These and other distinct characteristics of BLM lands reflect their curious history. Think back to Chapters Two and Three, in which we explained the initial acquisition, disposal, and withdrawal of federal public lands. Between 1781 and 1867, the federal government acquired nearly 2 billion acres of land and transferred over half of it into state and private ownership. Of the land it retained, much was placed into the National Park System, National Forest System, or National Wildlife Refuge System. Everything else, roughly speaking, is BLM land.

DOI: 10.4324/9781003392514-8

In short, BLM land is what remains of the original public domain. It is the land that Congress did not withdraw for specific purposes or transfer to state and private ownership. When Congress in 1976 declared its intention to retain permanent ownership of BLM lands, it froze the public domain lands in the configuration we see today. There are expanses of open desert, checkerboarded sections where the federal government owns every other square mile of land, and countless small, scattered tracts (often 40, 80, or 160 acres) surrounded by private or state land. And permanent federal ownership did not do away with most federal land and resource disposal laws. In some places, BLM lands are littered with mining claims. Elsewhere, fracking pads dot the BLM's landscape.

It may seem like BLM lands are leftovers or remnants. While there is some truth to this characterization, you shouldn't conclude that they have less value than other federal public lands. As this section will explain, BLM lands provide wildlife habitat every bit as essential as wildlife refuges, and some BLM lands can match national forests in their timber worth. The BLM's National Conservation Lands System includes areas that rival prominent national parks in their scenic and cultural significance. The BLM produces more money for the U.S. Treasury than any other federal land agency, since it administers all federal onshore mineral development—including not only oil, gas, and coal but also minerals that are crucial for renewable energy and battery technology. This, along with their suitability for wind and solar power installations, makes BLM lands a focal point for federal energy policy.

National Conservation Lands

Let's begin with BLM lands that have been given specific designations by Congress, the President, or the Interior Secretary. Thirty-six million acres of BLM land (about 15% of the total) have received some form of conservation status, and these together are called "National Conservation Lands." Most of these area designations—wilderness areas, wilderness study areas, wild and scenic rivers, national scenic trails, and so forth—are found within all four federal land systems. The BLM manages its conservation areas to the same standards as other agencies, with the notable exception that BLM units are more likely to allow livestock grazing even though it might seem at odds with the land's formal status.

The history of these designations reflects the relatively late appreciation for BLM lands as a permanent part of the federal estate. When Congress passed the Wilderness Act in 1964, it excluded BLM lands from wilderness review; only in 1976 did Congress broaden the Act to include BLM lands. And the BLM's first national monument was only designated in 1996.

Table 8.1 Designations of National Conservation Lands Managed by the BLM.

National Monuments	28
National Conservation Areas and Related Designations[1]	25
Wilderness Areas	260
Wilderness Study Areas	487
National Wild and Scenic Rivers	81
National Scenic Trails	5
National Historic Trails	13
National Wild and Scenic Rivers and Riverways	10
Conservation Lands of the California Desert	1

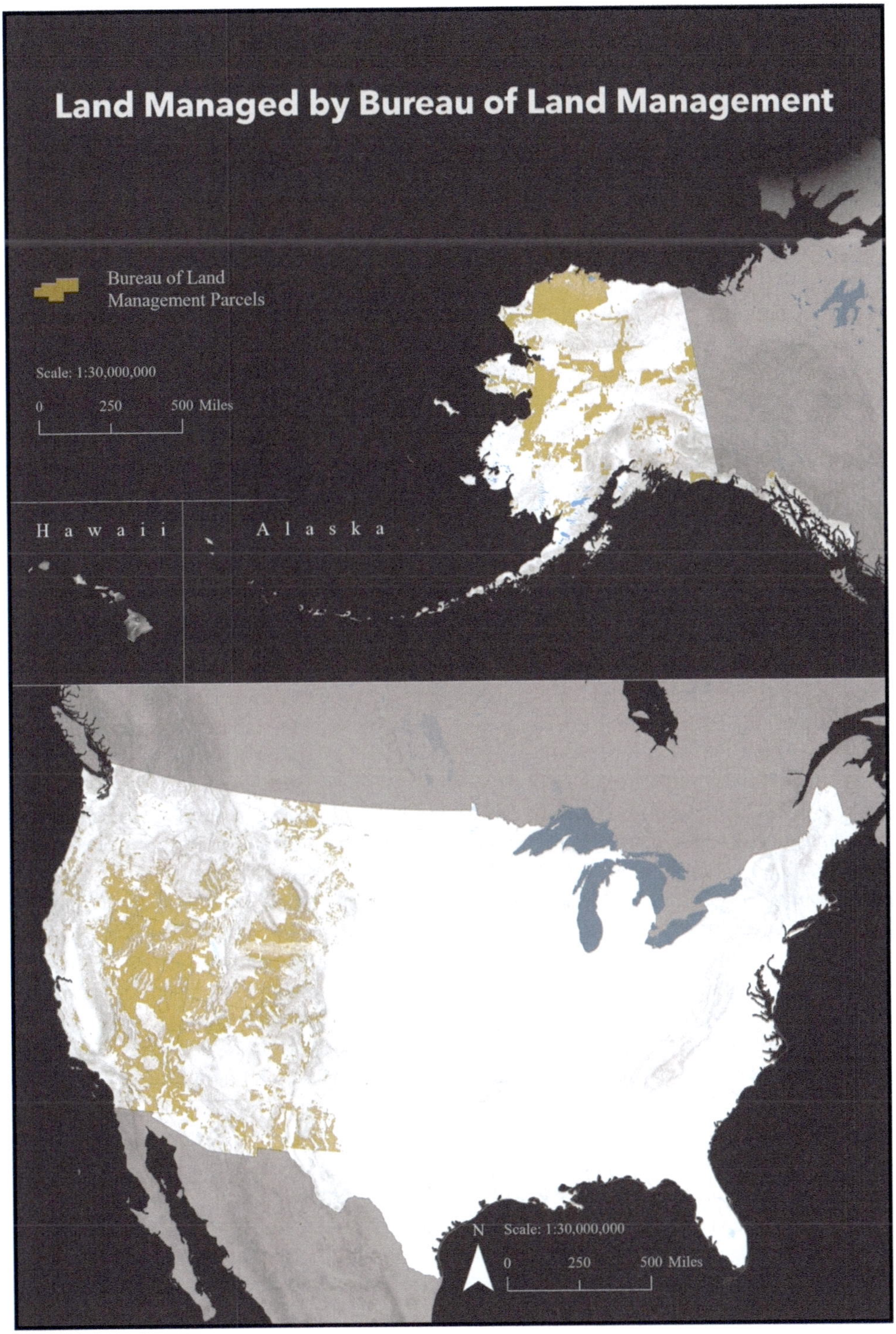

Figure 8.1 Land managed by the BLM. Map by Addison Arendt.

National Conservation Lands make up only a small percentage of BLM land, but in absolute terms they are substantial: their 36 million acres make them about as large as half the entire National Park System.

Oregon and California Revested Lands

The BLM manages a handful of other significant land areas with special resource emphases, including the Oregon and California Railroad Revested Lands (the so-called "O & C Lands") on the western side of the Cascade Mountains in Oregon. The O & C Lands total about 2.6 million acres given by the federal government to railroad and wagon companies to fund development but then revested (taken back) after the companies failed to meet the terms of the original grant.[2]

The O & C Lands are noteworthy because, unlike most BLM lands, they are covered by rich Douglas fir forests, including some of the last remnants of old growth forest in the entire Pacific Northwest. The management of old growth forest is a political lightning rod. Towering, ancient trees are lucrative for timber harvesters but also create an irreplaceable ecosystem. The O & C forests are essential habitat for the threatened Northern Spotted Owl and other vulnerable species. They provide other critical ecosystem services, such as water quality protection, and they have recreational and cultural value to millions of people. Not surprisingly, these lands have been the subject of intense political debate and litigation over the last 40 years.

In the Oregon and California Revested Lands Act of 1937, Congress mandated that the O & C lands were to be managed for sustained-yield timber production to support local economies, protect watersheds, and provide for outdoor recreation. In essence, Congress created a public forest system but assigned most of its management to the General Land Office and later the BLM instead of the Forest Service. In addition, Congress required the BLM to give 50% of all revenue from these lands to the counties in which they are located, making O & C management decisions critical to county budgets. The more timber the BLM sells, the more money counties receive for their budgets. That amount has declined precipitously in recent decades. In 1972, the BLM authorized the harvest of 1.2 billion board feet of timber from O & C lands. But shifts in management to protect vulnerable species and other ecological values have reduced timber harvesting considerably. For the last 20 years, the BLM has authorized only about 190 million board feet. The net result for Oregon counties containing O & C land has been a sharp drop in federal payments—from an average of $150 million between 1960 and 1993 to about $25 million in 2022.[3]

Since the early 1990s, Congress has compensated for this decrease by providing additional, direct payments to the affected counties. Meanwhile, county governments and the timber industry have fought to increase timber harvesting. While harvest levels may increase in the future, it is difficult to imagine that they will ever return to peak levels because the BLM is now required by law to balance timber harvesting with the protection of vulnerable species and other ecological values.

There is one last point of interest regarding the O & C Lands. As you learned in Chapter Two, land grants to railroads generally included alternating square miles of land along the railroad route. The O & C lands reflect this pattern, creating a checkerboard of land ownership in western Oregon. This presents myriad management challenges; among other things, it complicates road access and impairs wildlife habitat. Needless to say, the O & C Lands are likely to require disproportionate attention from BLM officials for some time to come.

Figure 8.2 Western Oregon, with the O & C Lands shown in brown. Source: The Spatial History Project, Stanford University.

The National Petroleum Reserve

Another significant area of BLM land, the National Petroleum Reserve—Alaska (NPRA), is located on the western end of Alaska's vast North Slope. Originally set aside as a naval petroleum reserve in 1923, this 23-million-acre area was transferred to the BLM in 1976. The U.S. Geological Survey currently estimates that the NPRA contains 8.7 billion barrels of technically recoverable oil (enough to supply the entire U.S. for over a year) and 25 trillion cubic feet of natural gas (just under a year's worth of U.S. consumption).

The NPRA was reserved by Congress for a dominant use: oil production. This constrains BLM's approach to land management, which ordinarily allows BLM to select among multiple land uses. Nevertheless, the NPRA has remained largely a reserve rather than an area of active production. The primary oil production area on Alaska's North Slope lies to the east, at Prudhoe Bay. But production there has decreased significantly in recent years, and as a result there is growing pressure to expand oil production in the NPRA.

Public perceptions of the NPRA have changed significantly since 1923, leading to renewed debate. Like the Arctic National Wildlife Refuge discussed in Chapter Six, the NPRA provides essential habitat for migratory birds and other wildlife. It is also utilized for subsistence hunting and fishing by Alaskan Natives, who have significant land holdings along the coast. But because Congress reserved the NPRA primarily for its energy resources, the real issue is not whether the BLM will lease oil and gas resources in the Reserve, but rather where leases will be offered and the extent of those leases.

Mineral Estate

This brings us squarely to the issue of mineral development. When we talk about federal public lands, we are typically talking about surface lands owned by the federal government and open to the public. The BLM manages 244 million acres of surface lands, but it is also responsible for *all* onshore federal subsurface mineral rights—rights which underlie approximately 700 million acres. Most of this "mineral estate"—about 643 million acres—lies below federal public lands. The remaining 57 million acres of subsurface mineral rights lie under state and private land (an arrangement called a

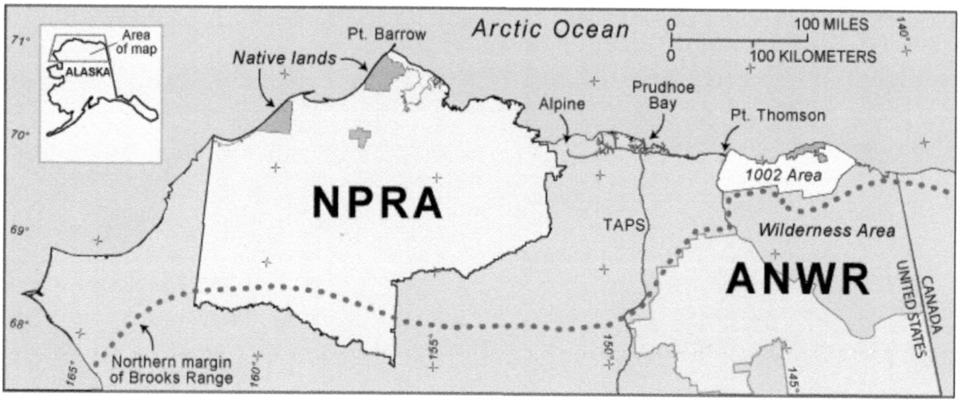

Figure 8.3 The National Petroleum Reserve in Northern Alaska. The Arctic National Wildlife Refuge (ANWR) lies to the east of the Reserve. Source: U.S. Geological Survey.

"split estate," in which surface and subsurface rights are owned by different parties). The BLM also has trust responsibilities for subsurface minerals under 56 million acres of surface land held by Indian tribes.[4]

The BLM's management of mineral development raises several difficult issues. First, where the BLM manages both the surface and subsurface, it must decide what areas to offer for development, recognizing that mining and drilling are incompatible with many other potential uses of the surface. For example, a decision to allow oil drilling may necessitate a reduction in grazing or recreational use. This is challenging enough. But even more vexing are those situations in which the BLM does *not* manage the surface above a planned mineral project. In these instances, the agency must negotiate and coordinate with the surface owner or manager, whether that be the Fish and Wildlife Service, the Forest Service, an Indian tribe, or a state or private landowner.

The BLM's Mission

Chapter Three explained that the BLM was formed in 1946 by merging the General Land Office, which was created in 1812, with the much younger U.S. Grazing Service, established by the Taylor Grazing Act of 1934. The BLM inherited their missions of land disposal, mineral leasing and patenting, surveying, and rancher-led range management. Through this consolidation, the BLM assumed responsibility for the remaining public domain lands, which were technically still slated for "final disposal," meaning transfer to state and private owners.

At its inception, then, the BLM was to be a real estate and resource agency, and BLM lands were viewed primarily as the raw materials for development. The agency's first seal, adopted in 1954, is essentially a vignette of the European-American frontier (see Figure 8.4). On the left, you see open rangeland with Conestoga wagons. In the middle you see the BLM, represented by a surveyor, a logger, an engineer, a cowboy,

Figure 8.4 The BLM's first seal. Source: U.S. BLM.

and a miner. The results of their work are depicted on the right: a railroad, oil derricks, smokestacks, and high-rise buildings. The BLM, in other words, was intended to advance industrial development.

That reputation persists, and environmental advocates often refer to the agency derisively as the "Bureau of Livestock and Mines." But the BLM has also evolved in interesting and important ways. In the mid-twentieth century, two developments changed the future of the BLM and the land it manages. First, the existing configuration of public domain laws proved inadequate to drive the transfer of remaining public domain lands to states or private owners. Second, public attitudes toward BLM lands shifted, putting pressure on Congress to integrate them into the permanent federal land base. Congress responded in 1976 with a sweeping declaration that the remaining public domain lands would stay in federal ownership permanently unless existing laws provided specifically for their transfer. Congress gave the BLM a multiple-use mandate very similar to that of the U.S. Forest Service: "protect the quality of scientific, scenic, historical, ecological, environmental, air and atmospheric, water resource, and archeological values … provide food and habitat for fish and wildlife … provide for outdoor recreation" while simultaneously promoting the development of minerals, timber, and forage.[5]

Like the Forest Service, the BLM's land planning process must contend with the myriad tensions embedded in its multiple-use mandate. But given the agency's constituents, funding, and staffing, the BLM has focused far less than the Forest Service on scientific research, visitor management, or ecological restoration. It is hard to say whether this will change in the coming decades. The BLM's "Public Lands Rule," finalized by the Biden Administration in 2024 but withdrawn by the Trump Administration the very next year, would have placed "conservation … on par with other uses of the public lands under FLPMA's multiple-use and sustained-yield framework."[6] This would have given BLM planners additional legal support for ecological restoration, fish and wildlife conservation, and climate adaptation, even when extractive uses were possible in a given planning area. But the rule did little to alter the agency's freedom to balance conservation with other uses, so whether such a rule could translate into policy change will remain unknown for the time being.

The Law and BLM Land Management

As noted above, the Federal Land Policy and Management Act of 1976 gave the Bureau of Land Management a multiple-use, sustained-yield mission almost identical to that of the U.S. Forest Service. The Supreme Court has described multiple use as "a deceptively simple term that describes the enormously complicated task of striking a balance among the many competing uses to which land can be put."[7] In Chapter Seven, you were introduced to the Forest Service's approach to multiple-use management, and you learned, for example, how the federal budget can shape the Service's year-to-year decisions. Similarly, Congress and the President guide the BLM via the budget and the articulation of presidential priorities. But the BLM is also bound by a variety of resource-specific statutes. This section reviews the most significant resource programs that the agency must address during its planning process. In addition to its forestry program on O & C Lands, there are major statutes governing mining, mineral leasing, range management, and wild horses and burros management.

Mining

The BLM's most significant mining laws are more than a century old. Several date back to the nineteenth century. Valuable hard rock minerals, such as gold and silver, are governed principally by the Mining Law of 1872, which was introduced in Chapter Two. This law authorizes prospectors to search for valuable minerals on unreserved public lands. Prospectors who stake a mining claim acquire the exclusive right of exploration within that claim. They can begin mining operations while title to the land is still held by the federal government. Miners pay no royalties on the minerals they produce. None. Zero. The minerals are simply theirs to keep in their entirety. In fact, until 1994, prospectors who met certain requirements could "patent" their claims, giving them outright title to the overlying land for $2.50 to $5.00 per acre, even if the land held millions of dollars' worth of gold, silver, or another mineral.

We'll delve more deeply into the Mining Law of 1872 in Chapter Thirteen. For now, the important point is that the Mining Law is fraught with management problems—how to decide between competing claimants, what to do about non-compliant miners, how to allow public access to and through mining claims—but its most controversial feature is its giveaway of land and minerals. Still today, the 1872 law gives billions of dollars in *publicly-owned* minerals to *private* mining companies for a pittance. This stands in stark contrast to FLPMA and other statutes, which require the federal government to sell public resources at fair market value. In one notable case in 1994, Interior Secretary Bruce Babbitt signed a patent for 1,800 acres near Elko, Nevada, to the American Barrick Resources Corporation for just $9,000 ($5 per acre). The land held a gold deposit worth an estimated $8 to $10 billion. Babbitt complained that it was "the biggest gold heist since the days of Butch Cassidy." The director of the Office of Management and Budget noted that the company paid "less than they charge (NBA) players for fighting during a game."[8]

In the face of these problems, Congress placed a moratorium on patenting mining claims in 1994. That moratorium is in place to this day, but broader reforms of the 1872 Mining Law remain out of reach. And the moratorium has not deterred prospectors from staking and developing new mining claims. The BLM reported in 2022 that there were 489,099 active mining claims on federal lands, totaling more than 11 million acres.[9] The Congressional Budget Office has estimated that these claims are producing about $1 billion in revenue each year, and the federal government receives nothing for it except for a few administrative fees.

Another major problem with mining on public lands is the recurrent failure of mining entities to clean up and reclaim mined land after their operations end. Public lands are dotted with thousands of abandoned mines, and many are causing significant environmental, health, and safety problems. Federal and state mining laws typically require companies to post bonds to ensure compliance with their reclamation rules, but this hasn't cleared the backlog of abandoned mines. There will be more about this too, in Chapter Thirteen.

Mineral Leasing

During the nineteenth century, "energy minerals"—coal, natural gas, and oil—could generally be obtained by claimants under ordinary land disposal laws, but the early twentieth century brought significant change. Between 1906 and 1916, Presidents

Roosevelt and Taft temporarily withdrew a total of 140 million acres of land from mineral entry, directed the U.S. Geological Survey to determine the existence and value of coal deposits on those lands, and replaced coal land disposal processes with a new leasing program. A similar approach followed for oil and gas resources, along with a program to maintain permanent oil reserves deemed vital for the U.S. Navy. President Roosevelt, in a 1907 special message to Congress, argued that

> the nation should retain its title to its fuel resources, and its right to supervise their development in the interest of the public as a whole … Mineral fuels, like the forests and navigable streams, should be treated as public utilities.[10]

After years of debate, Congress in 1920 passed the Mineral Leasing Act, which remains intact today. The law ended the disposal of land valuable for coal, oil, and gas and replaced it with a leasing system. While the Mining Law of 1872 allowed prospectors to enter and search for minerals almost anywhere on public lands, the Mineral Leasing Act gave the BLM's predecessor the authority to determine which parcels of public land to offer for lease. Energy companies could nominate tracts they believed valuable, but the federal government made the final decision. Under the leasing system, companies bid competitively for leases and pay royalties for all resources they extract. The royalty rate for oil and gas has historically been 12.5%, though there have been many efforts to increase it to match higher rates charged by states and private owners.

The BLM's mineral leasing program accounts for a significant part of the nation's energy production—about a tenth of all oil and gas production and a third of all coal production. In 2022, the BLM oversaw 283 coal leases on about 425,000 acres, and 37,496 oil and gas leases on about 26 million acres, representing over 96,000 oil and gas wells.[11] Energy production on this scale leaves an indelible footprint on BLM lands, and statistics like these have led to vocal criticism of mineral leasing. In the face of climate change and declining biodiversity, why should BLM lands remain open to oil, gas, and coal development? Why should the federal government charge so much less in royalties than state and private mineral owners? Why should the BLM offer new leases when many existing leases have not yet been developed? And when and why should energy development take priority over recreation, wildlife habitat, ecosystem health, and livestock grazing?

The Mineral Leasing Act contains a built-in counterweight to objections like these. Following a pattern that is now familiar, the Act mandates revenue sharing from mineral royalties. In the Lower 48 states, the BLM generally must give half of all its royalty revenue to the states in which the money is generated; in Alaska, it's 90%. As we have described elsewhere, this creates an incentive for states to encourage extraction and explains why states often balk when the BLM contemplates reductions to mineral leasing in order to reduce greenhouse gas emissions or protect threatened species.

Range Management

Though hotly contested in many parts of the West, livestock grazing is one of the oldest and widest uses of public domain lands. It long predates the BLM. In the nineteenth and early twentieth centuries, public domain lands were used freely by cattle ranchers and sheepherders. Indeed, in 1890 the Supreme Court wrote that

> there is an implied license, growing out of the custom of nearly a hundred years, that the public lands of the United States … shall be free to the people who seek to use them where they are left open and unenclosed, and no act of government forbids this use.[12]

At the same time, the Supreme Court also ruled that Congress had full authority to regulate public lands as it saw fit, and that its tacit approval of grazing in the past had not conferred any permanent rights to livestock operators:

> the Government has, with respect to its own lands, the rights of an ordinary proprietor to maintain its possession and to prosecute trespassers. It may deal with such lands precisely as a private individual may deal with his farming property.[13]

Congress did not flex its muscle on the issue of overgrazing, however, until the Taylor Grazing Act of 1934 (TGA). This Act was intended to reduce damage from overgrazing, to establish orderly use of the range, and to stabilize the livestock industry. The TGA authorized the BLM's predecessors to organize grazing districts made up of land deemed "chiefly valuable for grazing." The districts were subdivided into grazing allotments, which were allocated to livestock operators by way of 10-year renewable permits. Permittees were required to own land or water rights within or near the relevant district, and permits were issued on a "use it or lose it" basis—if an allotment went unused, the permittee risked losing their permit.

This basic framework for grazing on BLM lands remains in place today. But as in many other areas of public land policy, the environmental movement of the 1960s and 70s wrought considerable change. Under the TGA, BLM's range management program was run by livestock operators themselves through so-called "grazing advisory boards," which were predictably reluctant to implement restrictive changes. Thanks to FLPMA and other interventions, the program is now overseen by trained range conservationists, albeit with extensive input from livestock operators and the general public. This structural shift, however, has satisfied no one: the livestock industry feels chronically beleaguered, while environmentalists argue that range health continues to suffer. Many see the TGA's goals of orderly use and industry stability as a powerful source of inertia. It's likely the BLM will maintain domestic livestock grazing on public lands despite concerns about its ecological impacts.

The debate between grazing interests and their critics involves several recurring issues. First, and most basic, is the question of whether domestic livestock grazing should be permitted on the entire 155 million acres under BLM range management. Yes, all such land was once deemed "chiefly valuable for grazing," but public values have changed over time. Some environmentalists argue that livestock grazing should be reduced or terminated on many BLM lands, especially in hot desert areas, and that grasslands should be managed for wildlife and ecological services instead. Livestock operators, on the other hand, point out that they provide an essential public good in meat production, that the Taylor Grazing Act directs the BLM to support the livestock industry, and that livestock grazing is itself an important tool for range management.

A second point of friction is related to the first. Critics bemoan the sheer amount of grazing authorized each year. In 2023, nearly 18,000 grazing permits were in force on BLM lands, authorizing over 12 million animal unit months (AUMs) of grazing. (An AUM is defined as one month of grazing by one cow/calf pair, one horse, or five sheep or goats.)[14] Since the agency's formation, it has reduced the total number of authorized AUMs by more than 40% in order to maintain a sustained yield of forage and to protect range values beyond livestock. In some cases, reductions were necessary due to obvious overgrazing and soil erosion. In other cases, reductions reflect more subtle, shifting values in range management, such as a growing appreciation of the importance of riparian areas (land along waterways) for biodiversity. Looking ahead, climate change may reduce the overall biological productivity of desert range, leading to further reductions in grazing levels. Whatever their basis, reductions are a regular source of disagreement and are reflexively opposed by public land ranchers.

This leads directly to a third issue. Do livestock operators have a *right* to graze on public lands? The Supreme Court has ruled quite clearly that grazing permits are privileges, not rights, and that the BLM can revoke them. But practically and politically, the issue is much more complicated. Livestock owners usually own "base" property that is linked by law to their grazing permit. When the base property is sold, the grazing permit is generally transferred to the new owner, and appraisers include the value of grazing permits in their appraisals of ranch properties. So even though the permit is a privilege rather than a right, it functionally can be bought and sold with a ranch, and cancelling or reducing a grazing permit can significantly reduce the market value of a ranch. For this reason, even local banks resist reductions in grazing for fear that mortgaged real estate will be devalued.

Similarly, livestock operators often own water rights under state law, and retaining those rights generally requires that water be put to "beneficial use." Cancelling a grazing permit can make it difficult for a livestock operator to exercise their water rights and can therefore threaten those rights. Finally, ranchers often build range improvements—such as fences, holding pens, wells, and drinking troughs—on BLM lands but at their *own* expense. When grazing rights are diminished, those improvements often must be abandoned. For all these reasons, a large number of livestock operators, livestock associations, and western politicians insist that grazing permits are, in fact, property rights, despite the Supreme Court's ruling to the contrary.

A final point of regular controversy is the fee charged to ranchers for public land grazing. BLM permit holders pay just $1.35 per AUM, a rate that has persisted for decades and that is far lower than that charged by private landowners, the states, or other federal agencies. Officially, the formula used to generate that rate is designed to reflect the tension between two federal commitments—the mandate to earn fair market value for public resources, and the obligation to maintain the stability of the livestock industry. In practice, however, the latter commitment has prevailed: adjusted for inflation, the fee per AUM has declined considerably over the last 40 years.

Wild Horses and Burros

Many people think of domesticated animals and wildlife as entirely distinct categories, but somewhere between them are the tens of thousands of feral horses and burros that occupy BLM lands. These animals are not truly wild but descend from domestic animals that escaped onto public lands long ago. In the Wild and Free-Roaming Horses and Burros Act of 1971, Congress declared

> that wild free-roaming horses and burros are living symbols of the historic and pioneer species of the West; that they contribute to the diversity of life forms within

> the Nation and enrich the lives of the American people; and that these horses and burros are fast disappearing from the American scene.

Congress passed the Act to bring an end to the inhumane capture, harassment, and killing of these animals on federal lands.

The Act sailed through both chambers of Congress unanimously, reflecting strong public support. But wild herds have multiplied, making their management difficult and controversial. Today, the BLM estimates that there are almost 83,000 wild horses and burros on BLM land, which the agency says is two or three times the land's carrying capacity. Although the Act authorizes the BLM to euthanize excess wild horses and burros, this is a last resort both legally and politically. The agency sterilizes some animals, but its primary management approach is to round up excess animals humanely and put them up for public adoption. Since 1971, the BLM has adopted out some 290,000 horses and burros. Even so, supply perennially exceeds demand. As a result, the BLM is forced to care for thousands of captured animals, spending millions of dollars on its wild horse and burro program.

Regulations and Resource Management Planning

Laws governing forestry, mining, range management, and wild horses and burros deserve special attention, given their outsized impact on the agency. But these make up only a fraction of the agency's statutory authorizations, which take up about 1,150 pages in the U.S. Code. As with the other agencies, BLM translates statutes into management actions through regulations and resource management planning.

The BLM's regulations fill about 2,000 pages in Title 43 of the Code of Federal Regulations. These regulations are critical, most obviously because they represent the agency's official interpretation of federal statutes. But with Congress largely gridlocked on major land, resource, and environmental issues, significant changes in the BLM's legal directives today come about through the administrative rulemaking process rather than through new legislation. Needless to say, regulations created by Democratic administrations differ enormously from those created by Republican administrations. Like other agencies, the BLM must navigate shifting political winds and the competing rules and priorities they usher in.

The resource management planning process requires the BLM to translate its statutory and regulatory mandates into general management goals. FLPMA, after outlining the BLM's broad, multiple-use mission, obligates the BLM to "prepare and maintain on a continuing basis an inventory of all public lands and their … values." It also requires the BLM to "develop, maintain, and, when appropriate, revise land use plans."[15] The BLM therefore prepares a Resource Management Plan (RMP) for each resource area, spelling out how it intends to balance multiple uses for the next decade. It typically integrates the RMP with an Environmental Impact Statement (EIS) to comply with NEPA. The preparation of an RMP/EIS can be a massive administrative undertaking. One review found that it took the BLM an average of 2,252 days to go from a Notice of Intent to the final Record of Decision for an RMP/EIS. The process may require millions of taxpayer dollars and can be quite contentious.[16] When the BLM recently released a draft RMP in Wyoming, some state legislators attacked it as "probably the biggest disaster in the history of the United States." That was an obvious exaggeration, but a revealing one. In some areas, BLM plans are singularly important to the local economy.

Over time, RMPs have revealed a slow reduction in a variety of extractive land uses, including modest declines in grazing, oil and gas development, and off-road vehicles, and an increase in designated Areas of Critical Environmental Concern. There has also been an uptick in the amount of land devoted to the production of wind and solar power. This creates interesting tensions even within the environmental community because renewable power installations can require significant acreage and may occupy the habitat or migration routes of threatened species. Although the reasons for these various land use changes are multiple and complicated, they all suggest growing public interest in and concern over the ecological health of BLM lands. As always, however, trends can shift as quickly as the wind.

BLM Operations

Despite managing more land and generating more revenue than the other three land agencies, the BLM has the second smallest staff (10,320) and the smallest budget ($1.4 billion) among them. Indeed, the BLM manages three times as much land as the National Park Service but has less than half the funding. This is a policy choice by Congress, and while it no doubt reflects apolitical factors (such as demand for recreational amenities), many perceive in this choice a desire to tie the BLM's hands in deference to western communities and natural resource industries.

Organization

How are the BLM's operations different from the other federal land agencies? First, as we have already highlighted, the BLM is a Western agency; all but 30,000 acres of BLM land are located in eleven western states and Alaska. This has political implications, since congressional delegations from just twelve states have an outsized stake in the direction of BLM management. Regionalism is a major factor in political battles over BLM land. Several western states recurrently pass legislation either claiming or demanding ownership of BLM land. For the time being, such legislation is purely symbolic, but it does hint at some western states' dim view of the BLM.

BOX 8.1 BLM Headquarters as Proxy for Political Control

Do BLM lands belong to the nation or to the states in which they are located? Although the law's answer is clear—BLM lands are federal, not state—the question has long been contentious. In the Sagebrush Rebellion of 1979, many western states passed legislation challenging the legitimacy of federal public land ownership, focusing primarily on BLM lands. In wave after wave of legislation and litigation, western states like Nevada and Utah have continued to press this claim. The issue has become national and partisan; the Republican Party's national platform in 2016 and 2020 included a commitment to transfer BLM lands to the Western states.[17]

When President Trump first took office, he followed President Reagan's approach, which was to ignore the question of ownership and focus on the question of control, particularly control by conservative rural voters and natural resource industries. The Trump Administration's most significant symbolic gesture was to move the BLM's headquarters in 2020 from Washington, D.C. to Grand Junction, Colorado, creating administrative chaos. The administration's rhetoric was that of democratic populism:

western lands needed western headquarters so that agency leaders would be near the people they served. The BLM shared its new headquarters building with Chevron and the Colorado Oil and Gas Association, a lobbying firm. Choosing Grand Junction over Denver spoke volumes, since it put the BLM's headquarters in a rural Western town with no nonstop flights to Washington, D.C.

President Trump's chief of staff at the time, Mick Mulvaney, emphasized that the move would also weed out staff who had greater allegiance to Washington, D.C. than to the West itself. In this respect he was correct. More than 80% of BLM staff in the Washington office retired or resigned rather than move to Grand Junction.

A year and a half later, the Biden Administration announced that the BLM headquarters would return to Washington. The administration emphasized that even before the initial move, more than 95% of BLM employees nationwide lived in the West or Alaska, giving westerners easy access to decision makers. Interior Secretary Deb Haaland explained that "the BLM should have a leadership presence in Washington, D.C.—like all the other land management agencies—to ensure that it has access to the policy-, budget-, and decision-making levers to best carry out its mission."[18] The Biden Administration's decision portrayed BLM lands as national public lands rather than lands of merely regional importance.

Second, unlike the other land agencies, the BLM's administrative boundaries align with state boundaries (see Figure 8.5). This reflects the agency's distinct relationship with state governments. The BLM has taken steps to align itself with the Interior

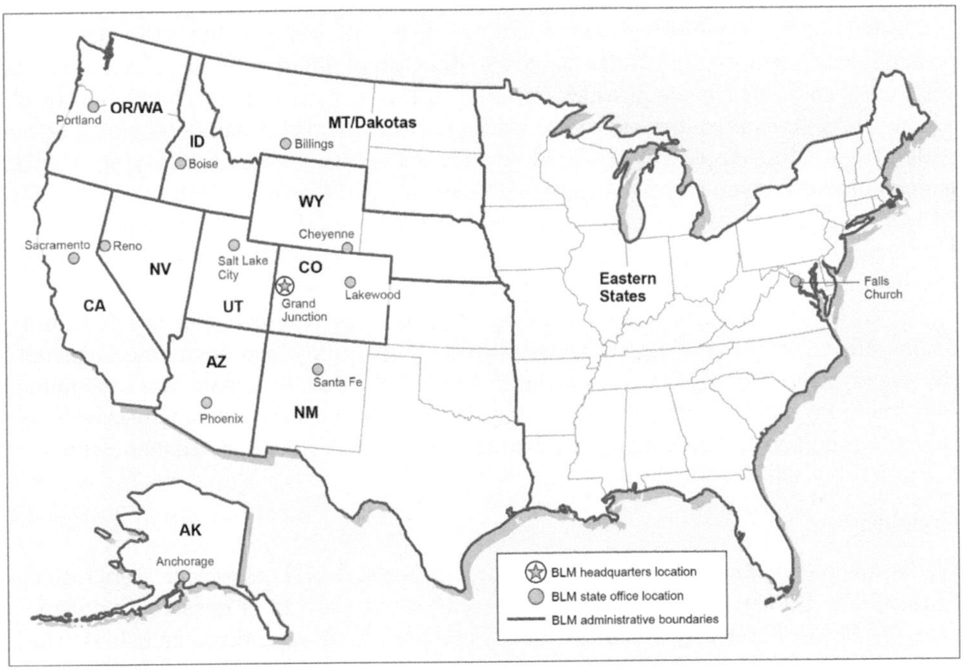

Figure 8.5 Locations and Administrative Boundaries of the twelve state BLM offices. Source: U.S. GAO.

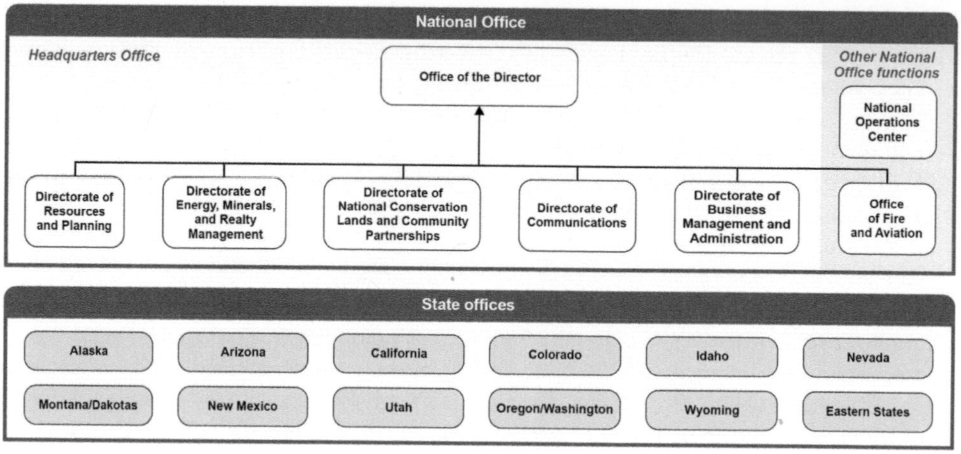

Figure 8.6 Bureau of Land Management Administrative Organization. Source: U.S. GAO.

Department's unified regional delineations, which are based on watersheds rather than state boundaries, but it is unlikely that this will change the dynamics of BLM state offices any time soon.

Organizationally, the BLM is led by a politically appointed director and two deputy directors. The agency's assistant directors oversee programs or administrative areas, which are identified in Figure 8.6. Notice the similarities and differences with other agencies. For example, the BLM has an assistant director for energy, minerals, and realty, who oversees the BLM's unique mineral management responsibilities as well as the many real estate transactions involving states and private parties. But the BLM does not have a position equivalent of the NPS's associate director of interpretation, education, and volunteers, since the agency provides far fewer visitor services than the NPS.

The BLM's overall staffing likewise reveals its emphasis on resource programs. More than 12% of BLM employees are hired into the geologist job series, and about 10% fall under range management and forestry. Another 4% work in realty and surveying. The BLM's staffing in biological sciences falls well behind the other land management agencies. Despite the agency's slow evolution, it is still funded and staffed to prioritize resource development, which it must balance with other values in multiple-use management.

The BLM has few resources for law enforcement—just 288 law enforcement officers, or one officer per 850,000 acres. These officers often work alone in remote areas, and the agency relies heavily on county sheriffs and other outside assistance. By comparison, the NPS has over 500 officers for its urban units and more than 1,000 law enforcement rangers in the national parks, which is about one officer per 50,000 acres.

Conclusion

We've now completed our survey of the four primary federal land agencies. Each one is distinct. The Bureau of Land Management stands out from the Fish and Wildlife Service, the Forest Service, and the National Park Service in several respects. It stewards the nation's public domain legacy, administering the federal land that was not claimed by settlers or granted to states. It emphasizes resource development over scientific research, visitor services, and ecological conservation—and thus finds itself doing business and battle with ranchers, timber harvesters, miners, and Western state and county governments. More than the other agencies, the BLM is squarely in the middle of national energy policy. Its lands are suitable for not only fossil fuel production but also renewable energy installations and mines for critical minerals. And thanks to the Bureau's lean staffing, BLM lands are among the least regulated and monitored public lands in the nation. For those who find national parks crowded or restrictive, the BLM offers an alternative.

Having examined these agencies in turn, we now shift our attention to a handful of the most acute public policy issues confronting public land managers today. The chapters that follow will take up these issues one at a time, explaining how they are addressed across the federal estate. As you read, remember the differences between the four agencies covered in Chapters Five through Eight. Each agency's ability to meet present challenges is shaped by the characteristics of the land it manages, the mission that guides it, the laws that constrain it, and the way the agency is organized and structured.

Study Questions

1 Explain the key conflicts arising from the BLM's multiple-use mandate and its balance with other values.
2 Why does the BLM receive smaller annual appropriations than the other federal land management agencies, and how does its budget and staffing affect its operational capabilities?
3 How does the Mineral Leasing Act of 1920 differ from the Mining Law of 1872 in terms of federal control, revenue generation, and disposition of resources?

Resources for Further Study

Muhn, James, and Hanson R. Stuart. *Opportunity and Challenge: The Story of BLM* (Washington, D.C.: U.S. Department of the Interior, 1988).
"Public Land Statistics 2023." Washington, D.C.: Bureau of Land Management, 2024.
Skillen, James R. *This Land Is My Land: Rebellion in the West* (New York: Oxford University Press, 2020).
Skillen, James R. *The Nation's Largest Landlord: The Bureau of Land Management in the American West* (Lawrence, KS: University Press of Kansas, 2009).
Vincent, Carol Hardy, *Land Disposal Authorities and Processes of the Bureau of Land Management* (Washington, D.C.: Congressional Research Service, 2024).

Notes

1 In addition to national conservation areas, the BLM manages six areas with related conservation purposes, such as "outstanding natural areas."
2 The BLM manages about 2.1 million of these acres, and the Forest Service manages the remaining 0.5 million.
3 Anne A. Riddle, "The Oregon and California Railroad Lands (O&C Lands): In Brief" (Washington, D.C.: Congressional Research Service, 2023).
4 Mariel J. Murray, "Energy Leasing and Agreement Authorities on Tribal Lands: In Brief" (Washington, D.C.: Congressional Research Service, 2023).
5 This language comes from the Federal Land Policy and Management Act of 1976 and is codified at 43 U.S.C. § 7301(a)(8).

6 89 Fed. Reg. 40308 (May 9, 2024).
7 *Norton v. Southern Utah Wilderness Alliance*, 542 U.S. 55, 58 (2004).
8 Jon Christensen, "Babbitt Attacks Mining's Gold Heists," *High Country News*, May 30, 1994.
9 "Public Land Statistics 2022" (Washington, D.C.: Bureau of Land Management, 2023), 132.
10 41 Cong. Rec. 2806 (1907).
11 "Report on the Federal Oil and Gas Leasing Program" (Washington, D.C.: U.S. Department of the Interior, 2021), 4.
12 *Buford* v. *Houtz*, 133 U.S. 320, 326 (1890).
13 *Camfield v. United States*, 167 U.S. 518, 524 (1897).
14 "Public Land Statistics 2023" (Washington, D.C.: Bureau of Land Management, 2024), 83.
15 43 U.S.C. § 35.
16 John C. Ruple and Mark Capone, "NEPA, FLPMA, and Impact Reduction: An Empirical Assessment of BLM Resource Management Planning in the Mountain West," *Environmental Law* 46 (2017): 953.
17 For a more complete history of this conflict, see James R. Skillen, *This Land Is My Land: Rebellion in the West* (New York: Oxford University Press, 2020).
18 Deb Haaland, "Press Release: Secretary Haaland Outlines Next Steps to Rebuild Bureau of Land Management" (Washington, D.C.: Department of Interior, September 17, 2021).

9 Climate Change and the Public Lands

Introduction

Climate change is the mother of all environmental problems. It shifts shorelines, strengthens hurricanes, causes raging floods and scorching droughts, and forces the migration of humans and animals. No corner of the earth is untouched by its effects. Although our understanding of climate change grows by the hour, humanity's efforts to mitigate climate change have mostly fallen flat. But that is a story for another book. This chapter examines only a tiny sliver of the climate crisis; namely, how it affects public lands and their management.

Given the scale of climate change, you might have expected this chapter to be lengthy. But lawmakers in the U.S. have not yet crafted a coherent body of law to address climate change. Without a climate-oriented legal mandate, federal agencies must instead incorporate climate concerns into their other activities. This means that other chapters in this book, while focused primarily on other topics, necessarily raise some of the climate-related efforts undertaken by land agencies. Chapter Eleven, for example, addresses biodiversity and discusses how climate change shapes agencies' efforts to comply with the Endangered Species Act. Rather than rehash those discussions, this chapter emphasizes three principal concerns: first, the political context of climate change; second, greenhouse gas emissions that emanate from public lands; and third, federal land managers' approach to adaptation in the face of climate change.

The Political Context of Climate Change

The issue of climate change is incredibly complicated. Tragically, however, the issue has been sucked into a political maelstrom, creating a polarized debate in which most Americans find themselves either strongly in favor of, or strongly opposed to, government efforts to combat climate change.[1] This polarization especially infects the U.S. presidency. Democratic presidents tend to promise robust climate action, while Republican presidents urge against it, with each successive presidency increasing in urgency and vitriol. Because presidents have significant authority over federal agencies, those agencies find themselves bouncing back and forth between opposing policies.

This yo-yo effect has profound consequences for federal land management. Chapter Four noted that federal law gives land managers a great deal of discretion over not only specific land use decisions but even how to implement legislation through regulation. This means that presidents are free to backtrack on their predecessors' commitments and to revise or reverse their regulatory initiatives. Aggressive presidential action, as exemplified

DOI: 10.4324/9781003392514-9

by President Trump's second term, sometimes takes the form of sweeping away long-standing rules, norms, and assumptions about how best to carry out the duties imposed by statutes like the National Environmental Policy Act (NEPA), the Endangered Species Act (ESA), and the Federal Land Policy and Management Act (FLPMA).

On matters related to climate change, this phenomenon has been extreme (see Table 9.1). For example, the Obama and Biden Administrations used existing statutory authority to effect significant new climate-related analyses and protections. Republicans objected both to these policies themselves and to the assertion of presidential authority underlying them—but then wasted no time, during the Trump presidency, in using that authority even more stridently. The second Trump Administration began with a furious escalation, purging federal rules and policies of any reference to climate change. The Department of Interior was instructed to prioritize oil, gas, and coal production and to abandon efforts to encourage renewables. NEPA processes in place since the 1970s were eliminated in a stroke, allowing extensive projects to be approved with barely a hint of environmental analysis. The ESA was reinterpreted to remove many protections for endangered species' habitat. And perhaps most disturbingly, the scientific infrastructure that helped build our understanding of climate change—federal grant programs, agency research programs, and the like—was systematically dismantled.

Table 9.1 A Sampling of Executive Orders Related to Climate Change and Public Land Management in the Biden and Trump Administrations.

President	*Order No.*	*Title*	*Date*	*Description*
Biden	13990	*Protecting Public Health and the Environment and Restoring Science to Tackle the Climate Crisis*	Jan. 20, 2021	Formalized a moratorium on oil and gas leasing in the Arctic National Wildlife Refuge; reinstated reliance on science-based decision-making across federal agencies.
	14008	*Tackling the Climate Crisis at Home and Abroad*	Jan. 27, 2021	Directed the Secretary of the Interior to pause new federal oil and gas leases on public lands; charged agencies with creating a Civilian Climate Corps to help restore and manage federal lands.
	14072	*Strengthening the Nation's Forests, Communities, and Local Economies*	Apr. 22, 2022	Initiated efforts to protect old-growth forests on federal lands, improve wildfire resilience, and expand reforestation to reduce carbon emissions.
Trump	14148	*Initial Rescissions of Harmful Executive Orders and Actions*	Jan. 20, 2025	Revoked several Biden-era Executive Orders related to climate change, including 13990 and 14008 (listed above).
	14153	*Unleashing Alaska's Extraordinary Resource Potential*	Jan. 20, 2025	Commanded a wide range of agency actions to accelerate resource development in Alaska.
	14156	*Declaring A National Energy Emergency*	Jan. 20, 2025	Declared a national energy emergency to expedite energy-related approvals (leasing, siting, production) on public lands.

Thanks to the structure of American law- and policy-making, climate policy is likely to oscillate between opposite poles for some time to come. Unless Congress acts in a dramatically new fashion, federal land managers will remain caught in the middle of a political firestorm. As you read the rest of this chapter, which discusses how these managers have dealt with climate change, bear in mind that, depending on when you are reading, many of their actions may have been dismantled by a hostile administration or enhanced by a supportive one. Only time will tell which policy shifts are enduring and which will survive only as long as the current president.

Emissions on the Federal Public Lands

In this section, we'll explore the most significant sources of carbon dioxide and methane emissions associated with the public lands. Anthropogenic (human-caused) climate change is the result of the increasing concentration of greenhouse gases (GHG) in the earth's atmosphere. Although carbon dioxide (CO_2) is the principal greenhouse gas, there are many others. Methane (CH_4) is critical because its warming potential is nearly twenty times that of CO_2. Even water vapor is considered a greenhouse gas because it can trap atmospheric heat. When quantifying GHG emissions, it's standard practice to liken them all to carbon dioxide—the typical unit of measure is one million metric tons of carbon dioxide equivalent (MMT CO_2 EQ)—but while this has the virtue of allowing different gases to be summed into one figure, it obscures the fact that different GHGs have different properties. For example, although methane is a more potent GHG than carbon dioxide, it breaks down much more rapidly: after roughly a decade in the atmosphere, oxidizing agents convert it into water vapor and carbon dioxide.

Greenhouse gas emissions result from many activities on public lands. Roads crisscross federal land, inviting drivers to travel long distances by car, or facilitating truck access to mining or extraction sites. Some public lands contain buildings and other development that requires considerable energy, much of which comes from fossil fuels. But by far the most important source of greenhouse gas emissions associated with the public lands is fossil fuel production.

Fossil Fuel Production on the Public Lands

As this book notes repeatedly, federal law has long allowed the extraction of coal, petroleum, and natural gas from the public lands. The extraction process itself causes some emissions, but these amounts are relatively trivial compared with the amounts caused by the combustion of fossil fuels at their point of use. In the aggregate, the CO_2 emissions from fossil fuels produced on the federal lands amount to roughly a quarter of *all* national GHG emissions.

Is it fair to attribute emissions to the public lands simply because the associated fuels originated on those lands? Whether it is fair or not, it is standard practice to link emissions *either* to the location of the emission *or* to the location where the relevant fossil fuel was extracted. Which of these two options is chosen depends on the purpose of the particular study. For our purposes, it's important simply to recognize that fuels produced on public lands are usually burned elsewhere—but public lands policy is still very much implicated in those emissions. At the same time, it's incorrect to assume that the U.S. could cut a quarter of its GHG emissions by banning fossil fuel extraction on public lands. Presumably, at least some (if not all) of that reduction would be offset by increased production elsewhere, whether private land, state-owned land, the offshore, or foreign states. Nonetheless, it is an important fact of federal land management that a large fraction of the U.S.'s GHG emissions are linked to the longstanding national policy of allowing commercial fossil fuel extraction on public lands. You'll learn more about that policy in Chapter Thirteen.

As between coal, oil, and natural gas, it is coal that has dominated federal land emissions in recent decades—and specifically coal mined in Wyoming, in areas like the Powder River Basin in the state's northeast corner. Most of the country's largest coal mines are concentrated in that region. As recently as 2014, nearly 70% of the entire carbon dioxide emissions associated with the extraction and combustion of fossil fuels produced on public lands derived from fuels produced in Wyoming, and the overwhelming majority of those emissions owe to coal from the Powder River Basin.[2]

Coal production nationally has been in decline. The dominant use of coal domestically is electric power generation, and recently the energy industry has been shifting away from coal and toward renewables and natural gas. This shift is driven both by steep drops in the price of energy from solar, wind, and natural gas, and by environmental regulation imposed on coal-fired power plants. The U.S. Energy Information Administration estimates that U.S. coal production dropped by over 23% between 2018 and 2023. If that trend continues, it will dramatically affect the GHG emissions profile of the public lands. On the other hand, crude oil production on the public lands has risen considerably at the same time, more than doubling between 2016 and 2021. By 2021 it represented just over 10% of all crude oil production in the United States. Natural gas production on the public lands rose about 1% during the same period. As you can infer, market dynamics—and specifically changes in demand for various fuels—can affect public land fuel production in sudden and unexpected ways.[3]

BOX 9.1 Do Presidential Politics Affect Fossil Fuel Extraction from Public Lands?

During presidential election campaigns, candidates and their political parties often make grandiose claims about fossil fuel production on public lands. Conservative candidates tend to champion fossil fuels (you may recall the chant, "Drill, baby, drill!"). Left-leaning candidates, by contrast, tend to emphasize the protection of public lands and the risks of climate change. Listening to these claims, one might assume that presidents wield direct control over the amount of fossil fuel extraction taking place on public lands.

The truth is considerably more nuanced. Fossil fuel production is shaped by market forces far more than by presidential intervention. The president lacks the authority to terminate existing oil and gas leases or mining operations; those operations are based on contractual or property rights that cannot legally be swept aside. Similarly, although the president can make more federal land available for fossil fuel production, she or he cannot command private businesses to purchase rights to such land. Whether to bid for resources on public lands is fundamentally a business decision.

For example, consider the perennial dispute about whether to expand fossil fuel production in the Arctic region of northern Alaska. During President Trump's first term, his administration made some of this land available for oil and gas leasing—but when leasing rights were offered at auction, bidding was tepid. The industry's lukewarm response to the auction resulted from both a soft market at the time the auction was held and from concerns that any rights obtained could be vulnerable to later changes in policy. In other words, the market responds not only to the policies of the current administration but to bidders' expectations about future administrations (and, indeed, the volatility created by the mere fact of presidential turnover).

Fugitive Emissions from Fossil Fuel Production

To this point, our discussion has focused on emissions from the combustion of fossil fuels. But fossil fuel production causes GHG emissions in other ways. Some emissions are linked to the process of fossil fuel extraction. These emissions are referred to generally as "fugitive emissions," and they are far more difficult to measure than emissions related to fuel combustion.

Let's use coal mining as an example. Coal is formed when layers of organic matter are compressed underground over millions of years. This same process creates methane, too, and indeed methane gas can often be found, in varying quantities, trapped within coal deposits. (This phenomenon gives us the phrase "canary in the coal mine." Methane gas is dangerously poisonous but odorless and invisible, so miners would send birds down underground mining shafts; a dead canary would indicate that methane levels were unsafe.) So ordinary coal mining activities involve—indeed, require—the venting of methane into the atmosphere. As noted, methane is a potent greenhouse gas. But assessing the quantity of methane embedded in a coal deposit is a tricky business. It is well demonstrated that methane continues to seep out of coal mines, whether surface mines or underground mines, even long after they have been abandoned. In recent years, technologies have emerged that help scientists quantify methane emissions by way of remote sensing capabilities—for example, by imagery taken by satellite or helicopter. But even our best estimates of fugitive emissions from coal mining remain imprecise compared to emissions from coal combustion.

Fugitive emissions associated with oil and gas production are similarly difficult to measure. Methane is often mixed underground with oil, just as with coal. So when a well is drilled into an oil or gas field, the resulting mixture of fuels often includes both liquids and gases. If there is no pipeline infrastructure to capture the methane gas—probably because the well owner is interested primarily in the crude oil for economic reasons—it must be vented or burned off (see Box 13.4, "Flaring Methane on the Public Lands," in Chapter Thirteen). And even if there is a pipeline available, no pipeline system is entirely free of leaks. It is well known that natural gas transportation systems leak between 0.5–2.5% of the fuel they carry. These too are fugitive emissions, and as you can imagine, measuring any of these emissions with precision is not easy. The net result is that for now our knowledge about emissions on the public lands is incomplete and imprecise. It's likely that the quality of our data will improve significantly in the coming years.[4]

Historically, federal law did not require federal land managers to address fugitive emissions (although such emissions, wherever they occur, may be subject to limited regulation by the U.S. Environmental Protection Agency (EPA) under the authority of the Clean Air Act). But especially during Democratic presidential administrations, the land management agencies have faced pressure to limit fugitive emissions from oil and gas production. In 2024, for example, the BLM issued a new regulation called the "Waste Prevention Rule." The Rule aims to reduce venting or flaring whenever technically and economically feasible. Producers are required to establish leak detection and repair programs and to pay royalties on natural gas that is "avoidably lost." Whether or not the Rule withstands legal or political challenges, it is a good example of a climate-aware regulatory initiative that is within the current authority of the BLM by virtue of its jurisdiction over oil and gas leasing on federal lands.

BOX 9.2 Geologic Carbon Sequestration on Public Lands

One approach to decarbonization involves capturing CO_2 from industrial sources, such as power plants or cement production facilities, and then injecting it (in liquefied form) into underground geological formations. This approach has not yet been deployed at a large scale, but the BLM has indicated that its land may be available for carbon sequestration projects of this kind. BLM land may be especially suitable for geologic sequestration for several reasons. First, sequestration projects typically use rock formations that previously contained oil or gas, and they rely on production infrastructure that remains in place. Because many oil and gas projects have been carried out on public lands, there is ample pore or reservoir space available, along with the wellbore equipment that can facilitate CO_2 injection. Second, because carbon sequestration projects involve large underground formations, they sometimes require the permission of numerous landowners overlying the formation. With its substantial landholdings, the BLM can simplify this process by making available large tracts of subsurface rights.

In recent years, policy and financial incentives have accelerated the potential for carbon sequestration on public lands. BLM has provided guidance for issuing rights-of-way for carbon dioxide sequestration projects, providing a clearer regulatory pathway, and Congress has enacted tax credits to make these capital-intensive projects more economically viable. But significant questions remain. Scientists are uncertain whether sequestration techniques will remain secure over long time periods, and devising ways to monitor stored CO_2 is no easy task. There is some evidence that sequestration may induce seismic activity. Liability for leakage or seismicity is unclear. Ultimately, the economic viability of large-scale pipeline infrastructure, as well as public acceptance, will be critical factors in determining the future of geologic sequestration on federal lands.

Greenhouse Gas Emissions and Public Lands Ecosystems

Discussions about anthropogenic climate change appropriately focus on human activities. But natural processes affect GHG emissions, too. Grasses, plants, and trees are constantly absorbing CO_2 from the atmosphere and converting it into carbon by way of photosynthesis. This is a natural form of carbon sequestration, and it is happening all the time. Some of that carbon returns to the atmosphere when those organisms respire while they are alive, decompose as they die, or burn in a wildfire. These natural fluxes can't easily be measured with exactitude, but they can be estimated, and they represent a very important part of the overall emissions profile of the public lands. The U.S. Geological Survey recently estimated that natural carbon sequestration by terrestrial ecosystems on the federal public lands offsets approximately 15% of the CO_2 emissions resulting from fossil fuel production on the public lands.[5]

BOX 9.3 Managing Carbon Stored in Alaskan Public Lands

Of all the nation's territory, Alaska is the most consequential carbon "sink." (A carbon sink is a geographical feature that absorbs more carbon from the atmosphere than it releases.) Alaska represents 18% of the U.S. landmass but more than 50% of its embedded carbon stock. This carbon is stored in vegetation, soils, and sediments, and it is especially abundant in the Arctic tundra. Some of the carbon flux associated with Alaskan public lands cannot be directly controlled by land managers. Melting areas of permafrost, for example, release methane into the atmosphere, but most such melting results from temperature changes over which federal agencies have no control. But other sources of flux can be mitigated by sound land management practices. Wildfires, for example, transfer significant quantities of carbon from forests into the atmosphere. Although much wildfire activity eludes management (see Chapter Ten), there are measures that can limit the incidence of wildfire and moderate the loss of forest carbon that would otherwise result.

Beyond wildfire, the enormous carbon stores in Alaska's permafrost and soils face significant risks from resource extraction and associated infrastructure development. The BLM's National Petroleum Reserve-Alaska, for example, overlays a massive expanse of carbon-rich tundra. Exploring and drilling for oil and gas and constructing roads, pipelines, and drilling pads can physically destroy tundra vegetation and the underlying soil where carbon is sequestered. Furthermore, these activities can alter local hydrology, leading to landscape-level changes that can accelerate permafrost thaw and subsequent carbon release.

The boreal forests of interior Alaska, which are largely under federal jurisdiction, are another critical component of the state's carbon inventory. These forests are susceptible to large-scale insect outbreaks, which are becoming more frequent and severe with warming temperatures. Infestations of species like the spruce bark beetle can kill vast tracts of trees, transforming a thriving forest from a net carbon sink into a significant carbon source. The dead, dry timber left in the wake of these outbreaks also dramatically increases the risk of wildfire, creating a feedback loop of carbon release. Insect infestations can be combated, but forest managers would need extraordinary resources and personnel to make a dent over such a vast territory.

In short, decisions made by federal land managers in Alaska will have a profound impact on the future of the state's immense carbon repository.

Importantly, land management decisions can take advantage of the natural carbon cycle to shape public lands emissions. For example, a manager's decision to allow a young forest to mature can sequester additional carbon within that forest. Harvesting mature timber, or allowing it to burn in a fire, can result in emissions. Of course, there are numerous factors that bear on a manager's decision to allow a timber harvest or to schedule a controlled burn in a forest. Although federal law currently does not require agencies to maximize carbon sequestration, GHG fluxes are becoming more salient to federal land managers. More generally, federal land managers increasingly regard GHG tabulations as a critical aspect of NEPA analysis (see Box 9.4)—though this is, as always, subject to the rapidly-changing political winds discussed at the start of this chapter.

BOX 9.4 NEPA and Carbon Emissions on Public Lands

As explained in Chapter Four, the National Environmental Policy Act (NEPA) requires a "detailed statement" about the environmental impacts of "major Federal actions significantly affecting the quality of the human environment." For years, federal agencies wrestled with NEPA's applicability to carbon emissions. Many argued that because climate change is a global problem playing out over a long timeframe, no single federal project's carbon emissions will "significantly affect the quality of the human environment." But several federal courts ruled otherwise, requiring land managers to report the carbon emissions associated with projects on federal land.[6]

Eventually, the agencies adopted the practice of estimating even downstream emissions—that is, emissions that would occur off the federal lands but due to projects conducted on federal land. For example, a NEPA filing for a coal mining permit issued by BLM would include a computation of the carbon to be emitted when that coal was burned in a power plant. Some agencies went a long step further and reported the "social cost" of those emissions, using a controversial standardized metric known as the "Social Cost of Carbon." This metric was intended to reflect the total societal price tag of the effects of a unit of carbon emissions.

The second Trump Administration, however, adopted a radically different view of NEPA's requirements and withdrew longstanding rules about NEPA implementation, eliminating any requirement to report carbon emissions or their social cost. Gone also was any computation of the global impacts of domestic projects. NEPA analysis was shortened and streamlined in myriad ways, weakening the statute's application to climate change. Ultimately, the federal courts are likely to play a sizeable role in determining what NEPA requires in the context of carbon emissions.[7]

Renewable Energy on Public Lands

Of course, federal lands can also be used to mitigate GHG emissions elsewhere. One of the major ways this can be accomplished is by taking advantage of lands suitable for renewable energy development. Wind and solar installations tend to require large tracts, and assembling land across multiple private owners can be difficult. Large, contiguous federal landholdings can offer a more straightforward path to development.

The BLM in particular has permitted a number of large renewable projects. Significant portions of its land possess excellent solar, wind, and even geothermal energy potential, often in remote areas with few existing conflicts. The flexibility inherent in its multiple-use mission makes it a natural fit for accommodating new energy projects, as compared to the more narrowly defined mandates of the Park Service or Fish and Wildlife Service. Finally, the agency has been involved in energy development for decades. This experience provides the agency with managerial frameworks, staff expertise, and administrative processes that can be adapted for renewable energy projects. The agency has also facilitated the construction of transmission infrastructure critical for connecting renewable projects to the grid. There will be further discussion of renewable development on federal public land in Chapter Thirteen.

Climate Change Adaptation on the Public Lands

As the climate changes, public land managers must adapt their management strategies to conform to new and emerging realities. Recognizing that some degree of climate change is unavoidable, all four federal land management agencies have developed strategies—in fits and starts, and in very different ways—to adapt to its impacts, aiming to reduce vulnerabilities and enhance the resilience of the resources and values they manage. Before we delve into these strategies, however, we must first discuss the climate science that undergirds the agencies' responses to climate change.

Climate Science and Adaptation

Climate science is a massive, sprawling enterprise. To comprehend the causes, consequences, and complexities of climate change requires the collective expertise of myriad disciplines; to solve the problem of climate change would require a truly godlike capability. So it is no surprise that the federal land management agencies struggle to know how best to respond to the risks of climate change. The science of mitigation is difficult, but at least it has the virtue of being tractable: it is relatively easy to compute the carbon emissions associated with, for example, an oil and gas lease or even a forest fire. By comparison, the science of adaptation can seem virtually impenetrable.[8]

This impenetrability stems from several factors. Many climate change impacts are highly localized and context-dependent. A slight increase in temperature might have negligible effects in one ecosystem but devastating consequences in another. Predicting these nuanced, site-specific impacts is exceedingly difficult—indeed, in many instances it is beyond the reach of current scientific techniques. Even if an agency is confident about a particular climatic variation, it requires an enormous leap to understand precisely how that variation will affect specific species, migration patterns, hydrological cycles, vegetation, and much else. This complexity is compounded exponentially by the interconnectedness of ecological systems because a change in one component can trigger a cascade of unforeseen effects throughout an entire ecosystem.

To manage this complexity, climate scientists rely on models designed to forecast future conditions; land managers, in turn, must use these models to develop plans for adaptation. Deciding whether to take a particular action—to invest in drought-resistant species, relocate infrastructure, or implement a controlled burn—relies on modeled assumptions about how severe future droughts will be, how often wildfires will occur, or how much sea levels will rise. But climate models are built on probabilities, not certainties. So agency decision-making must incorporate not only an understanding of climate science but also of social and tactical considerations. What are the risks of getting it wrong? When making decisions under deep uncertainty, agencies must weigh the costs and benefits of various adaptive measures against the possibility that the predicted future may not materialize, or that unforeseen climate impacts will emerge.

On a more practical level, the temporal and spatial scales of adaptation often clash with agency planning cycles and political realities. Ecological adaptation can take decades or even centuries, far exceeding typical budget cycles or political terms of office. Implementing a long-term strategy, such as assisted migration of species or large-scale restoration projects, requires sustained commitment and funding that can be difficult to secure even in the best of times.

All these factors are complicated by the politics, both national and local. Adaptation measures can be controversial, particularly if they involve significant changes to land use, impacts on local communities, or perceived threats to traditional practices. Navigating these scientific, logistical, and socio-political complexities makes developing and implementing effective climate change adaptation strategies a monumental task for federal land managers. Almost always, the "right" decision is far from clear. Although we'd like climate science to point clearly to a particular course of action, simplistic judgments often reflect presuppositions of the proponent rather than an honest examination of the science of adaptation.

Vulnerability Assessment

Given this deep uncertainty, how do federal land management agencies actually handle adaptation? Although the agencies differ in their capacities and approaches, there are commonalities. First among them is the concept of *vulnerability assessment*. Each agency conducts assessments to understand which of its lands and resources are most susceptible to the effects of climate change. As discussed above, scientific evidence is not always available, and when it is, its implications are not always clear—but the goal of these assessments is to utilize the best science available, and especially an understanding of ecosystem sensitivities, in order to identify risks and inform adaptation planning and prioritization. Vulnerability assessment may be conducted regionally, or it may focus instead on a particular type of resource (such as coastal resources or a certain plant community).

These vulnerability assessments are critical for moving beyond general concerns about climate change to specific, actionable strategies. They typically involve several steps. First, the agency clarifies the scope and scale of the assessment. Next, agency staff gather and analyze relevant climate data and ecological information, looking especially for exposure to climate stressors such as increased temperatures or altered precipitation patterns. Then the sensitivity of land resources to these stressors is assessed, including, finally, the adaptive capacity of both natural systems and management strategies. The output is usually a detailed report that represents the agency's "best guess" as to the areas or resources at highest risk *and* the likelihood that the agency can manage the risk successfully. This systematic approach allows agencies to prioritize scarce resources and develop targeted interventions while looking with clear eyes at the limitations of the relevant science and the agency's capabilities.

Although all four of the agencies conduct vulnerability assessments, the Forest Service has probably done the most to incorporate such assessments into its core management functions. It generally focuses its vulnerability assessments on forest health and wildfire risk. In the Southwest, for example, the Service has conducted extensive assessments on the vulnerability of ponderosa pine forests to increased temperatures, prolonged droughts, and more frequent and intense wildfires. These assessments combine climate models with historical fire data, forest inventory data, and ecological knowledge of tree stress responses. The agency looks for specific forest stands or regions with high fuel loads, stressed tree populations, and limited adaptive capacity, which are then prioritized for fuel reduction treatments, prescribed burns, and reforestation with more drought-tolerant species (for more details about wildfire management, see Chapter Ten). Similarly, in the Pacific Northwest, the Forest Service has assessed the vulnerability of high-elevation forests to changes in snowpack and alpine hydrology. These assessments inform decisions on watershed management and species selection for future plantings.[9]

The National Park Service's approach to vulnerability assessment reflects its dual commitments to human enjoyment and resource preservation. This is evident, for example, in its assessment of coastal resources in units like Everglades National Park. Given its low elevation and proximity to the coast, the Everglades is highly vulnerable to sea-level rise and increased storm intensity. The NPS has conducted detailed assessments that integrate projections for sea-level rise with hydrological models and ecological data on species distribution. This effort identified critical habitats, such as mangrove forests and freshwater marshes, that are at high risk of saltwater intrusion and inundation. These findings led directly to adaptation strategies such as the restoration of natural water flow, the identification of potential areas for marsh migration, and the protection of visitor infrastructure in threatened areas.[10]

Like the National Park System, the National Wildlife Refuge System includes many coastal resources. For these units, the Fish and Wildlife Service focuses its vulnerability assessments on the exposure of tidal marshes, beaches, and migratory bird stopover sites to sea-level rise, increased storm surge, and changes in salinity. The Service integrates tidal models, elevation data, and species-specific habitat requirements to predict habitat loss and fragmentation. This information then guides decisions on coastal resilience projects, such as living shoreline installations, marsh migration easements, and the strategic placement of nesting islands.

Managing for Resilience

A second concept common to each agency's approach to adaptation is *resilience*. There is a clear and growing emphasis on the capacity of ecological systems to absorb disturbances, to reorganize after disruption, and to maintain essential functions and structures in the face of change and uncertainty. This may sound obvious, but for bureaucratized federal agencies, this emphasis involves a paradigm shift. Many land managers have, for generations, assumed that their objective was to maintain or restore ecosystems to some historical idea—often a vague sense of what a particular place should be like without human intervention. This approach has much to commend it and much to criticize, but whatever its merits in the abstract, the simple fact is that it may be unsustainable under climate change. The natural world is changing, and trying to replicate dated conditions will likely become increasingly impractical and idealistic. With resilience as the goal, agencies can devote their energies to fostering systems that can better cope with ongoing change.

A good example of this paradigm shift involves, again, the coastal units of the National Wildlife Refuge System, which are on the front lines of sea-level rise and increased storm intensity. Historically, refuge managers might have focused on maintaining specific wetland types through dikes and other infrastructure. However, with resilience as the guiding principle, some refuges are now undertaking "living shoreline" projects and identifying "migration corridors" for coastal habitats. At Blackwater National Wildlife Refuge in Maryland, for instance, managers have experimented with oyster reefs and natural marsh restoration techniques to dissipate wave energy and allow marshes to naturally migrate inland as sea levels rise. This approach prioritizes the capacity of the ecosystem to adapt to change, rather than attempting to hold a rigid line against the encroaching sea.[11]

Similarly, the National Park Service has altered its management strategies to promote resilience in the face of species migration and habitat shifts. As temperatures rise in high-elevation ecosystems, species like the whitebark pine, featured in several Western national parks (Crater Lake, for example), are jeopardized by warmer temperatures that inhibit cone production and facilitate the spread of mountain pine beetles. (The species was listed as "threatened" under the Endangered Species Act in 2023.) Instead of artificially maintaining historical whitebark pine distributions, the NPS has explored strategies such as protecting genetic diversity within the species, facilitating the assisted migration of the tree to more climatically suitable areas within park boundaries, or managing adjacent lower-elevation forests to provide stepping-stones for migrating species. The goal is to ensure the continued ecological integrity and functionality of these alpine environments, even if their species composition shifts over time.[12]

Adaptive Management

Finally, a third concept is *adaptive management*. Just as resilience is essential for natural systems, so also must the agencies themselves be willing to adapt to unfolding realities. "Adaptive management" may sound like a buzzword, but any human organization can become attached to an outmoded process; adaptive management simply reflects the idea that agencies should implement management actions as experiments, monitor outcomes, learn from the results, and then adjust their strategies iteratively over time. It's a straightforward idea, but you may be surprised at how difficult it has been for bureaucratic agencies to make even simple changes.

While all four agencies are converging around the central concept of resilience, their primary adaptation frameworks and implementation tools differ, reflecting their core missions and the nature of the resources they manage. The National Park Service has developed a simple decision rubric called the Resist-Accept-Direct (RAD) framework, which guides managers to decide whether to *resist* climate-driven changes (namely, to maintain historical conditions), to *accept* the transformation, or actively *direct* the system toward a new, desired state better suited to future conditions.[13] The agency also uses scenario-based planning to explore adaptation options under different plausible climate futures. The Forest Service does something similar; its Resistance, Resilience, and Transition (RRT) framework is comparable to RAD but emphasizes resilience-building silvicultural practices within existing forest ecosystems.[14]

These tools, like every other aspect of climate policy, are likely to come and go as agencies evolve and, especially, as presidents rotate through the White House. But most of today's land and resource managers are being trained to employ adaptive management as a matter of course, whether in bureaucratic, professional, or educational contexts. It's likely that iterative and experimental decision making will continue to become more prevalent, independent of the party in control of Congress or the presidency. As climate impacts become unavoidable, day-to-day activities are likely to require some amount of trial and error, and adaptive management will simply become standard operative procedure.

Conclusion

Climate change presents an unparalleled challenge to the management of U.S. public lands, fundamentally altering ecosystems and demanding a dynamic response from federal agencies. The sharply-divided political landscape exacerbates this challenge, causing climate policies to shift dramatically with each presidential administration. This instability makes long-term, consistent climate mitigation and adaptation efforts incredibly difficult for land managers caught in the middle of a political firestorm.

Furthermore, public lands are significant contributors to greenhouse gas emissions, primarily through the extraction and combustion of fossil fuels. While market forces play a substantial role in dictating fossil fuel production on these lands, federal policy historically has facilitated extensive extraction. Even with a national decline in coal production, the rise in crude oil extraction on public lands highlights the interplay of policy and market dynamics. The challenge is compounded by the difficulty in accurately measuring fugitive emissions from extraction processes and the natural carbon fluxes within ecosystems.

In response, federal land management agencies are increasingly focusing on adaptation, at least during sympathetic presidential administrations. These agencies employ vulnerability assessments to identify at-risk lands and resources, utilizing the best available science despite inherent uncertainties. Recognizing that maintaining historical conditions may no longer be feasible, the agencies are in the midst of a critical paradigm shift toward managing for resilience, fostering ecosystems that can absorb disturbances and adapt to ongoing change. Adaptive management, which treats management actions as experiments and encourages iterative adjustments based on monitored outcomes, is becoming crucial for agencies navigating these unfolding realities. The future effectiveness of these strategies will depend not only on scientific advancements but also on the ability of agencies to secure sustained commitment and funding amidst persistent political volatility.

Study Questions

1 Explain the difference between the traditional goal of maintaining historical conditions and the emerging management focus on ecosystem resilience.
2 Define the process of adaptive management and explain why it is crucial for agencies navigating the uncertainties of climate change impacts.
3 How does the polarized political context affect the continuity and implementation of federal land management policies that bear on climate change?

Resources for Further Study

Forest Service, U.S. Department of Agriculture. *USDA Forest Service Climate Adaptation Plan* (Washington, D.C.: U.S. Forest Service, 2022).

Klyza, Christopher McGrory. "Climate Change and the Management of Federal and State-Owned Land in the United States," in Gregory K. Ingram and Yu-Hung Hong (eds.), *Climate Change and Land Policies* (Cambridge, MA: Lincoln Institute of Land Policy).

National Park Service, U.S. Department of the Interior. *National Park Service Climate Change Response Strategy 2023 Update* (Washington, D.C.: National Park Service, 2023).

Sutherland, Sara. "The National Environmental Policy Act and Climate Change Adaptation Within Federal Natural Resource Management Agencies." *Case Western Reserve Law Review* 75, no. 1 (2024): 233–262.

U.S. Geological Survey, U.S. Department of the Interior. *Federal Lands Greenhouse Gas Emissions and Sequestration in the United States—Estimates for 2005–22* (Washington, D.C.: U.S. Geological Survey, 2024).

Notes

1 For a wise, enduring book about why climate change is so divisive, see Mike Hulme, *Why We Disagree About Climate Change: Understanding Controversy, Inaction and Opportunity* (Cambridge, U.K.: Cambridge University Press, 2009).

2 See, e.g., U.S. Energy Information Administration, "Sixteen Mines in the Powder River Basin Produce 43% of U.S. Coal," August 26, 2019.

3 See generally U.S. Energy Information Administration, *Annual Energy Outlook 2025* (Washington, D.C.: U.S. EIA, April 2025).

4 For official estimates of such emissions on the U.S. public lands, see U.S. Geological Survey, U.S. Department of the Interior. *Federal Lands Greenhouse Gas Emissions and Sequestration in the United States—Estimates for 2005–22* (Washington, D.C.: U.S. Geological Survey, 2024).

5 U.S. Geological Survey, U.S. Department of the Interior. *Federal Lands Greenhouse Gas Emissions and Sequestration in the United States—Estimates for 2005–22* (Washington, D.C.: U.S. Geological Survey, 2024).

6 See, e.g., *WildEarth Guardians v. U.S. Bureau of Land Management*, 870 F.3d 1222 (10th Cir., 2017).

7 In 2025, the U.S. Supreme Court issued a decision likely to diminish NEPA reporting of impacts not directly attributable to the permitted project, such as downstream GHG emissions. See *Seven County Infrastructure Coalition v. Eagle County, Colorado*, 605 U.S. 168 (2025).

8 For a discussion of the disconnect between science and land management, see Elaine M. Brice et al., "Impacts of climate change on multiple use management of Bureau of Land Management land in the Intermountain West, USA," *Ecosphere* 11, no. 11 (2020): e03286.

9 Janine R. Rice et al., *Climate Change Vulnerability Assessment of Aquatic and Terrestrial Ecosystems in the U.S. Forest Service Rocky Mountain Region* (U.S. Forest Service, 2018).

10 Peek, K.M., H.L. Thompson, B.R. Tormey, and R.S. Young. *Coastal Hazards & Sea-Level Rise Asset Vulnerability Assessment for Everglades National Park: Summary of Results* (NPS 160/186783, Program for the Study of Developed Shorelines, Western Carolina University, Cullowhee, N.C., 2022).

11 Georgetown Climate Center, *Managing the Retreat from Rising Seas: Blackwater National Wildlife Refuge, Maryland: Blackwater 2100* (Georgetown University Law Center, 2020).

12 National Park Service, *Bipartisan Infrastructure Law and Inflation Reduction Act support range-wide efforts to rescue disease-addled whitebark pine forests* (www.nps.gov, last updated Aug. 16, 2024).

13 National Park Service, U.S. Department of the Interior. *National Park Service Climate Change Response Strategy 2023 Update* (Washington, D.C.: National Park Service, 2023).

14 Forest Service, U.S. Department of Agriculture. *USDA Forest Service Climate Adaptation Plan* (Washington, D.C.: U.S. Forest Service, 2022).

10 Wildfire and the Public Lands

Introduction

Wildfire has received growing public attention in the last 20 years. With alarming frequency, the national media shows images of residential neighborhoods sitting helplessly before 150-foot walls of flame. The media's focus is primarily on large forest fires in the West. These catastrophic fires burn tens to hundreds of thousands of acres, cost millions of dollars to suppress, cause billions of dollars in damages, and take human lives. Most of these episodes involve public lands, forcing land management agencies to devote ever-increasing resources to wildfire management.

This chapter will explain how wildfire has grown from an unremarkable nuisance into a vast, urgent problem. It will discuss the causes and extent of the current wildfire crisis before turning to examine how federal land managers have responded. Led by the U.S. Forest Service, the federal government has experimented with fire suppression, controlled and prescribed burns, and fuel reduction strategies. The results remain ambiguous—but what is unambiguous is that federal expenditures related to fire management are rocketing upwards. The chapter closes by outlining the many challenges that lie ahead as federal land managers adjust to our "new normal" in which enormous wildfires, sad to say, play an ever-expanding part.

The Wildfire Crisis

Controlled fire is essential to human flourishing. We use it for space heating, cooking, transportation, industrial processes, and even ambiance. Indeed, most of the energy on which society depends is harnessed through fire, albeit fire hidden from view inside engines and generators. A *wild*fire, by contrast, is an unplanned fire burning in a natural or wildland area, such as a forest, grassland, or prairie. (That's why the terms "wildfire" and "wildland fire" are interchangeable.) Wildfire is a necessary but destructive ecological process. Where we want forest and grassland ecosystems, rather than urban areas or cultivated fields, we *need* fire. But all too often, wildfire is disastrous.

Catastrophic wildfire takes lives, destroys property, and scorches landscapes; these effects are well known and occupy the media's attention but are limited to a fire's locale. Wildfire smoke, however, is fast becoming a more widespread concern. In recent years, major cities like San Francisco have been brought to a virtual standstill by smoke from faraway fires. Wildfires in Western Canada now regularly cause haze in Chicago, Toronto, and even New York City. Smoke contains particulate matter and other unhealthy substances as well as vast amounts of carbon dioxide. The increase in annual

DOI: 10.4324/9781003392514-10

carbon emissions from wildfire negates many of the gains achieved by greenhouse-gas emissions reduction policies. In 2023, Canada's wildfires yielded four times the carbon emissions of the *global* aviation sector.[1]

Wildfire statistics can be staggering. In 2021, for example, the Dixie Fire became the largest single fire in California history. Over a period of 103 days, the fire burned nearly a million acres in the Plumas and Lassen National Forests and Lassen Volcanic National Park. It destroyed 1,311 structures and racked up a suppression cost of $637 million. That year, wildfire burned 7.1 million acres—an area larger than Massachusetts—and the federal government alone spent $4.4 billion dollars on fire suppression. But the full economic cost of wildfire is far higher even than that. The Congressional Budget Office puts it between $37 and $150 billion annually, depending on what the estimate includes.[2] By comparison, the Office's estimate of losses from hurricane winds and storm-related floods totals around $54 billion annually.

While the Dixie Fire is an extreme example, it is part of a disturbing trend. Since the National Interagency Fire Center began tracking wildfire in the mid-1980s, the average annual number of wildfires on federal, state, and private land has decreased, but the acreage burned has *doubled.* If we narrow our focus to federal lands, the statistics are just as concerning (see Figure 10.1). Although the average number of fires dropped by almost 20% between the 2000s and the 2010s, the average number of acres burned increased by 17% during the same period. That means that the average fire size increased by more than 40%.[3]

Of course, looking at the largest forest fires and the aggregate national data hides significant variation throughout the country. The Northeast region has the least wildfire activity. The relatively cool, wet climate limits the size and intensity of fires, and most fires are extinguished in one or two days. In the Southeast, the warmer climate

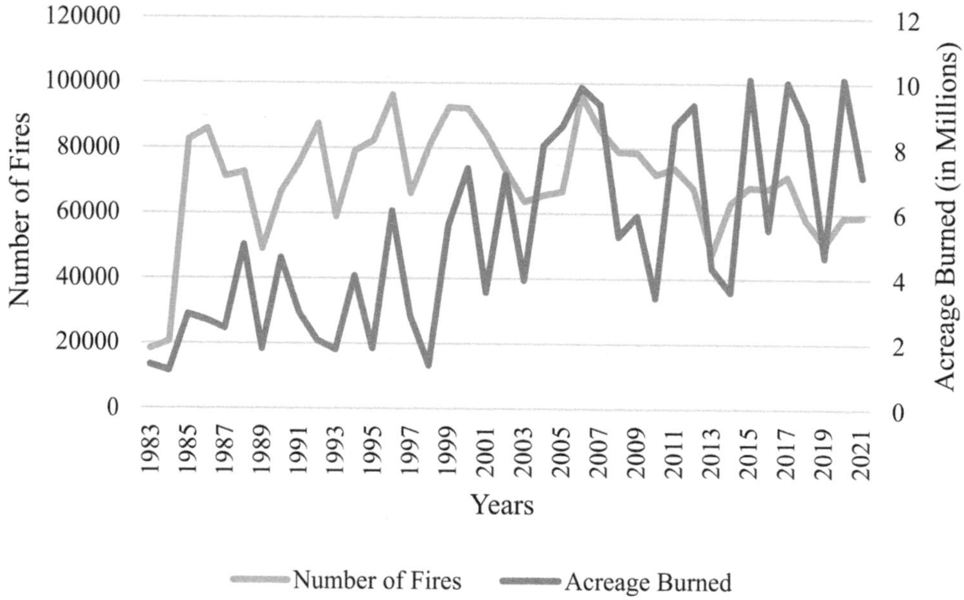

Figure 10.1 Annual Number of Wildfires and Acreage Burned on Federal Land. Based on data from National Interagency Fire Center.

and year-round fire season translates into significantly more wildfires. But the West receives by far the most attention, with the largest wildfires and most acres burned each year. Much of the region is arid or semi-arid, which constrains the urban and agricultural development that tends to limit wildfire. Furthermore, roughly half of the Western region is in federal ownership, characterized by vast expanses of unbroken forest and grassland. Much of the region has experienced intense drought over the last two decades, increasing its combustibility. Figure 10.2 maps the wildfire risk levels for federal lands.

There are also notable differences among federal land systems when it comes to wildfire (Table 10.1). Out of an annual average of 4.1 million acres burned on federal land from 2012 to 2021, 97% were under the jurisdiction of the Forest Service and the BLM.

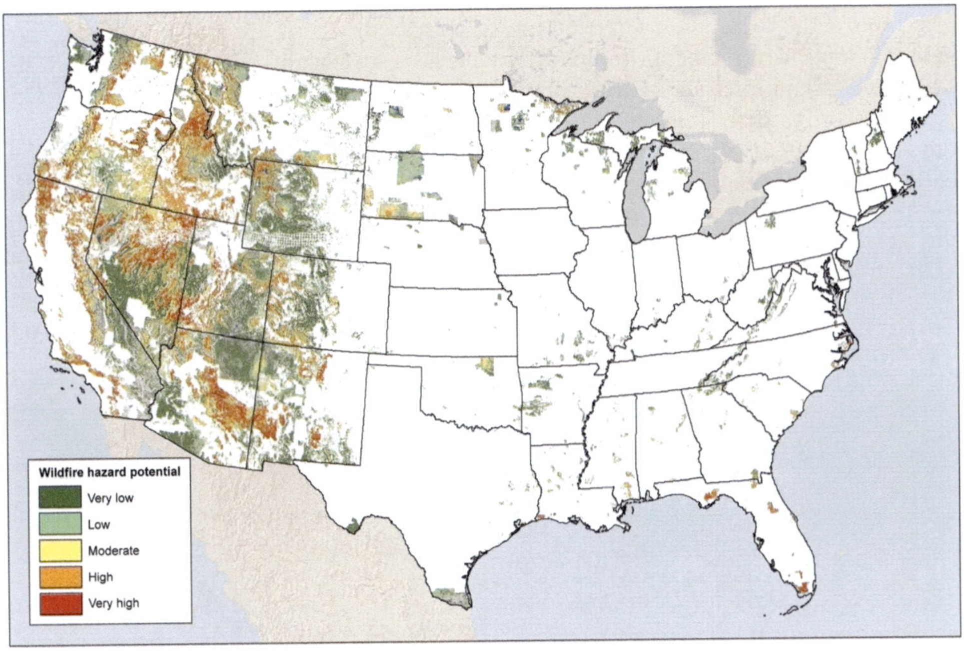

Figure 10.2 Wildfire Hazard Potential on Federal Public Lands in the Contiguous United States. Source: U.S. GAO.[4]

Table 10.1 Average Annual Wildfire Statistics by Agency, 2012–2021.

Agency	*Total Acres Managed (Millions)*	*Average Annual Fires*	*Average Annual Acres Burned*	*Percentage of Total Burned Acreage*
USFS	193	6,467	2,041,565	49
BLM	244	2,481	1,990,158	48
FWS	89	271	89,966	2
NPS	80	379	49,255	1

How Did the Wildfire Crisis Arise?

So how and why has wildfire become a crisis? The risk has increased for many reasons, but this chapter will focus on three. The first has to do with real estate development. Increasingly, people are choosing to live and build structures in fire-prone landscapes, a trend which has put more people and property at risk. The second reason is the buildup of fuel. In the context of wildfires, "fuel" refers to dry and combustible vegetation. Ironically, wildfire fuel increased across the West precisely because fire *suppression* was so successful for so long. Finally, wildfire risk is increasing due to climate change. The effects of climate change are uneven, but much evidence suggests that in certain parts of the United States it will make wildfire worse rather than better. Let's examine these three reasons in greater detail.

A Development Problem: The Wildland–Urban Interface

The most significant factor contributing to increased loss of life and property is burgeoning land development in fire-prone places—specifically, in what is called the "wildland–urban interface" (WUI, pronounced "WOO-ee" by those in the know). This is the jargon used to describe places where buildings are adjacent to forests or combustible vegetation. As the West has grown in population, more and more people have chosen to live in high-risk areas, perched on fire-prone hillsides or alongside parched forests. People are drawn to these areas for a variety of reasons. Many simply wish to live near nature, surrounded by trees or grassland. Others appreciate that rural

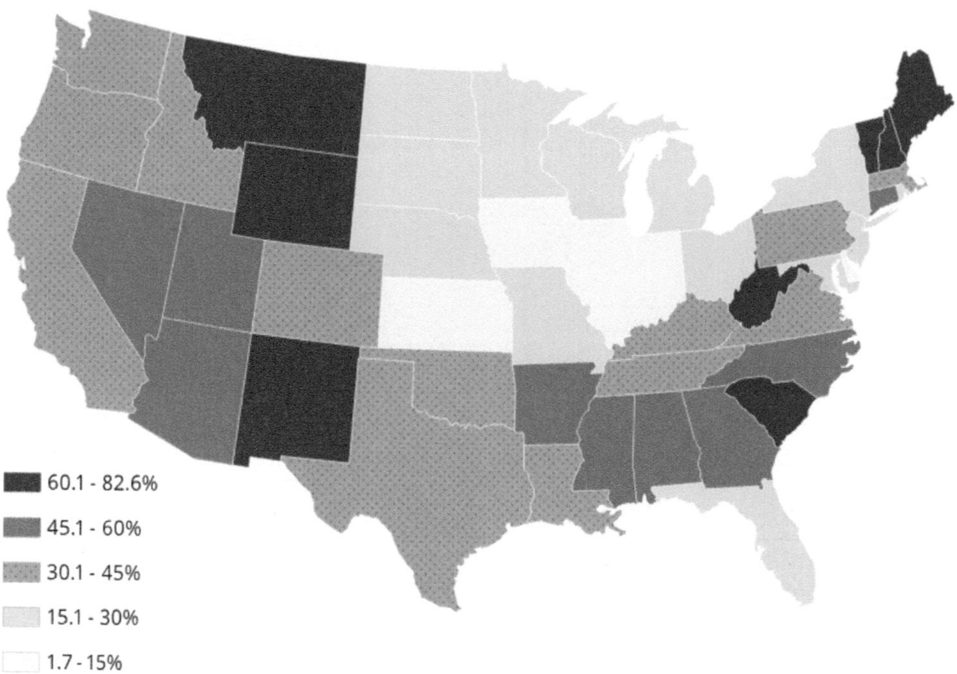

Figure 10.3 Houses in the wildland–urban interface as a percentage of total houses in the state. Source: U.S. Forest Service.

property is less expensive than urban property, or that rural zoning and building codes are less restrictive. But whatever the reason, from 1990 to 2010, the number of homes built in the WUI increased by 41%.[5]

Of course, the same trees that are scenic amenities today might be a tinderbox tomorrow. As people build structures in fire-prone landscapes, they dramatically increase the potential for loss related to fire. For example, the Hanley Fire in 1964 burned roughly 53,000 acres in Napa and Sonoma counties in Northern California. Since the area was sparsely populated, no one died, and only 108 homes burned. In 2017, the Tubbs Fire burned a much smaller area of 37,000 acres, the vast majority of which were within the original perimeter of the Hanley Fire. But because of 50 years of WUI development in that area, the Tubbs fire burned 4,651 homes and killed 22 people.[6]

To compound matters, human development increases the likelihood of *large* wildfires. Humans start most wildfires, including large wildfires, so growing populations and growing WUI development are associated with higher fire risk. Fires that begin in the WUI are also a major challenge for federal land management; most wildfires on federal lands start on adjacent private or state land. Having human neighbors *increases* fire risk for national forests, parks, and wildlife refuges.[7]

Development in the WUI has a significant impact on wildfire management. Fires in remote, uninhabited places require relatively little human intervention and can sometimes even be allowed to burn themselves out. But when human lives or significant properties are at risk, firefighters work aggressively to combat and contain wildfire. Not only does this require rapid and forceful action; it also means more firefighters working in more dangerous situations. This urgency also leads to skyrocketing financial costs, since agencies are under pressure to deploy more firefighters and to utilize expensive tools like air tankers and fire retardant. As one federal report stated,

> [i]n some western areas, the government pays more in suppressing fires than the fair market value of the structures threatened by those fires. It would literally be cheaper to let fires burn and pay 100% of the rebuilding costs.[8]

It might indeed be cheaper and also safer to let some structures burn—but this is simply not accepted practice.

A Fuel Problem

In some parts of the United States, the wildfire crisis is one we have brought upon ourselves, at least in part, as human modifications of the environment have made ecosystems less resilient. One example of this arises, paradoxically, from our success in fighting fires. Effective fire suppression in the past has increased the risk of catastrophic fire today by increasing the amount of unburnt fuel on the forest floor. During the first two-thirds of the twentieth century, the U.S. Forest Service made wildfire suppression a top priority. In 1935, the agency pledged to extinguish every fire by 10 a.m. the morning after it was reported, and in the next few decades, it enjoyed remarkable success in reducing the total number of acres that burned on the national forests each year.

Thanks to the Service's achievements, the fire regime in Western federal forests shifted dramatically. Ponderosa pine forests in the southwest had historically burned frequently but at low intensity, which kept the understory clear and favored mature trees that could withstand fire. With successful fire suppression, vegetation and woody debris built up on the ground. As a result, fires were less frequent but burned much hotter than in the past, leading to greater tree mortality and even soil sterilization. Accumulation of excess fuel has contributed to many of the largest fires of the past 20 years.

The other human modification that encourages wildfire has to do with invasive plant species. BLM land, which is predominantly grassland rather than forest, has experienced a significant uptick in the spread of non-native grasses and shrubs. Combustible invasive species, including cheatgrass, have expanded across the western grasslands, increasing fire frequency. To make matters worse, many of these invasives are pioneer species that thrive on recently disturbed sites, so more fire actually leads to more invasives.

A Climate Problem

Compounding these problems, climate change is increasing wildfire risks in a handful of ways. First, warmer conditions increase evaporation rates, drying out vegetation and making it more susceptible to ignition. Prolonged periods of drought are also becoming more common and severe in many regions. The combination of heat and diminished moisture is a recipe for wildfire. Similarly, the snowpack that normally accumulates in mountainous regions during the winter is a crucial source of moisture for the spring and summer. When more precipitation falls as rain rather than snow, and snow melts earlier in the year, this reduces the amount of water available to vegetation during the hot summer months, further contributing to dry conditions. These factors extend the fire season, lengthening the overall window of opportunity for wildfires to occur. Finally, the factors above also create conditions favorable for certain insect populations, such as bark beetles. These insects can infest and kill large areas of trees, leaving behind a significant amount of dead and dry fuel that can easily ignite and contribute to wildfire spread.

Managing Wildfire

As much as we'd like a silver bullet, there is no single solution to wildfire on or near federal lands. Federal agencies must continually experiment with a mix of at least four distinct approaches. They can suppress fire, excluding it from the land. They can let natural fires burn without suppression, allowing this ecological process to do the work it has accomplished for millennia. They can deliberately ignite a "prescribed fire" under favorable conditions to achieve management objectives. Finally, they can change a landscape's combustibility mechanically, usually by physically removing fuel from a forest.[9]

This section will discuss these approaches. But note, too, that in all cases, wildfire management must be adaptive, such that each set of proposals is treated as a hypothesis and tested over time.

And federal land managers must collaborate with adjacent landowners, lest their efforts be undermined by a careless neighbor. They must also collaborate with state, municipal, and private partners to support meaningful changes beyond their jurisdiction, such as building codes or zoning ordinances that mitigate the risk of fire.

Fire Suppression

Suppression is the first tool that federal agencies wielded against wildland fire more than a century ago. The Forest Service argued in 1905 that wildland fire suppression was probably the single largest public benefit of the national forests. A series of catastrophic wildland fires in the early 1900s only hardened the agency's resolve. The most dramatic was the Big Blowup of 1910, in which a breakout of fire on multiple national forests in Idaho and Montana, stoked by gale-force winds, created a firestorm that swept across the region on August 20. More than 3 million acres of federal and private forest land burned, destroying small towns, killing 85 people, and causing the loss of 7.5 billion board feet of merchantable timber.[10]

Chastened by events like these, the Forest Service embraced fire suppression with passion. The New Deal provided the agency with a significant influx of labor and funding through the Civilian Conservation Corps (CCC), Works Progress Administration (WPA), and the Emergency Conservation Work (WCW) programs. After World War II, the Service enjoyed new resources in the form of military surplus equipment and veterans eager for work. In 1945, the agency revealed its new fire mascot, Smokey the Bear, who reminded Americans that "Only You Can Prevent Forest Fires" (see Figure 10.4).

Measured in terms of acreage burned, the Forest Service's suppression program was a remarkable success. The federal government did not maintain consistent, detailed figures of total wildfire acreage before the 1980s, but general agreement exists that the number of acres burned each year dropped dramatically. In retrospect, this was a

Figure 10.4 Smokey the Bear, the popular emblem of the Forest Service's fire suppression efforts.
Source: National Agricultural Library.

pyrrhic victory in many forests, particularly in the West, because as we have seen, successful fire suppression allowed a dangerous degree of fuel accumulation. Today, suppression remains a valuable tactic, but it must be combined with fuel reduction efforts, which we'll discuss shortly.

How does federal fire suppression work? Responsibility begins at the local agency level. National parks, wildlife refuges, and forests, along with BLM district offices, have fire personnel on staff, and they are typically the first responders to a new ignition. But when local resources aren't enough, federal wildland fire resources are organized through regional coordination centers and, ultimately, the National Interagency Fire Center (NIFC) in Boise, Idaho. The regional centers have an equipment supply cache and can coordinate support from federal and state agencies, as well as private contractors, who provide aircraft, heavy equipment, food, and other supplies. The NIFC maintains yet more equipment, but more importantly, it can coordinate nationally across agencies to move resources where they are needed most.

BOX 10.1 Fighting Fire From the Air

A critical weapon against wildfire today is the DC-10 air tanker. Formerly a passenger aircraft capable of carrying 350 passengers, its previous purpose is still evident from the long row of windows down each side. But now a large tank hangs from the bottom of the fuselage. Inside the tank is 9,400 gallons of fire retardant, which can be dropped

Figure 10.5 A DC-10 air tanker dropping chemical fire retardant. Source: U.S. Forest Service.

in just four seconds, creating a fire line almost a mile long, at a price of about $70,000. The DC-10 isn't even the largest fire tanker. From 2015 to 2021, a company called Global Supertanker offered a converted Boeing 747 that could carry 17,500 gallons of fire retardant—roughly 114 times more capacity than the first air tanker, a Boeing Stearman 75 crop duster retrofitted in 1955 with a 170-gallon water tank.[11]

While almost all the staff involved in wildland fire suppression are civilians, suppression work often looks like a military operation. The logistics of deploying and supplying firefighters and support staff has few federal parallels outside the Department of Defense and the Federal Emergency Management Administration (FEMA). The challenges of coordinating billions of dollars of staff and equipment, on a moment's notice, are intimidating. The heavy equipment involved is substantial, from aircraft and bulldozers to mobile kitchens and communications centers. As many have observed, wildland fire suppression looks a lot like war, but in this kind of war, casualties are not acceptable.

Wildland firefighting is directed on the ground by an Incident Command System (ICS). The Forest Service and California officials developed the ICS in the 1970s for cross-jurisdictional wildland fires; the model is now standard across all federal, state, and local fire agencies.[12] The ICS creates a temporary command-and-control structure that can be constituted by qualified staff from any agency. The incidents are ranked from the simplest—Type 5—to the most complex—Type 1—and the ICS team's organization is scaled to that ranking. A Type 5 or Type 4 incident does not trigger the full ICS system and is handled by an Incident Commander (IC), utilizing local resources. By contrast, Type 2 and Type 1 incidents trigger the full ICS structure. For each incident, a pre-established ICS team is deployed to the fire. The team members include an IC as well as leaders for four functional areas: operations, planning, logistics, and finance (ee Figure 10.6). A Type 2 or 1 incident can include a wide range of other leadership and support positions as needed. These fires will have a dedicated public information officer to ensure communication to the public, and they will likely have branch leaders for things like aviation, food, and cost accounting.

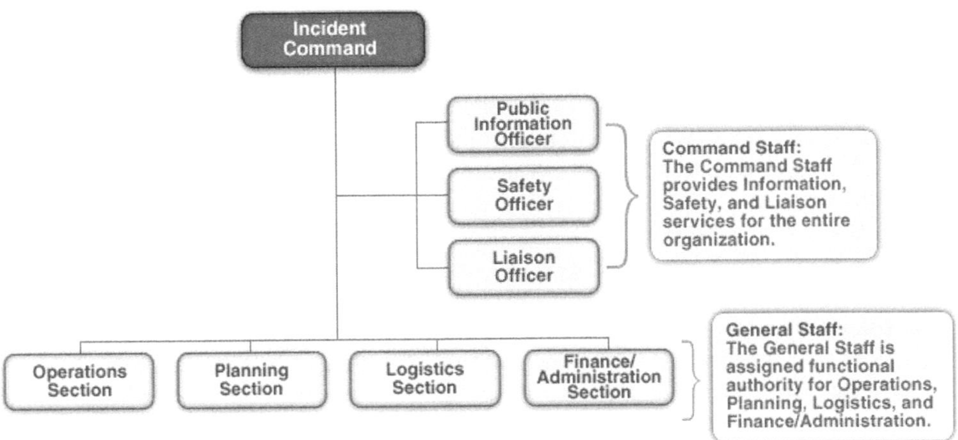

Figure 10.6 Incident Command System structure. Source: U.S. Forest Service.

The efficiency of the ICS system stems from the fact that team members have day jobs, such as park supervisors and district rangers, and are only mobilized when needed. The standardized training and certification process means that teams can be assembled from across agencies, and replacements can be found if needed.

In an active deployment, wildland firefighters cut fire lines with hand tools on the flanks of active fires, meaning they cut and scrape a pathway down to bear mineral soil that cannot burn. They also light "back burns" to deprive the fire of fuel and "mop up" the fire after it is contained by extinguishing remaining hot spots. There is a hierarchy of hand crews based on experience, training, and capabilities. Most firefighters are on Type 2 crews, typically of 20 people each. They reach fires in vehicles, rely on hand tools, and generally lack training for incident command or other specialized roles. Most people get their start in wildland firefighting on Type 2 crews. But federal land agencies also maintain about 100 Interagency Hotshot Crews, or Type 1 crews, staffed by experienced firefighters who have more training and credentials. They serve as first response teams for significant fires. Finally, the Forest Service and the Department of the Interior maintain nine elite "smokejumper" crews. These are the paratroopers of wildland fire, parachuting from planes to reach remote fires in roadless areas or places where it would take other crews hours or days to respond. Regardless of type, all of these crews can be deployed on fires for up to 21 days without a break, and they can work up to 16 hours a day.

While firefighting infrastructure is coordinated across the agencies, the burden of suppression is not equally shared among them. Most fire on federal lands is under the jurisdiction of the Forest Service. Of 18,700 federal wildland firefighters, 12,831 work for the Forest Service. More than half of those remaining work for the BLM.[13]

Managed Fire

In most American landscapes, fire has been a key ecological process. Indeed, some species cannot survive without fire. For example, several conifer species throughout the country have what are called serotinous cones. The cones are sealed tightly with resin and remain hanging in the tree long after their seeds mature. They open primarily when fire melts the resin. Conifers with serotinous cones include the Jack pine in the Northeast and Great Lakes regions, the Table Mountain pine in the Appalachian Mountains, and the Lodgepole pine throughout the West. Moreover, healthy forests require fire frequently enough to keep fuel (woody debris as well as dead and live standing trees) from building up. Fires are critical in grasslands as well. Fire on long and short grass prairies have historically kept forests from encroaching, and fire in sagebrush steppe habitat in the Great Basin has been essential in maintaining adequate forage for wildlife.

Fire itself, then, is an essential tool in land management. Even when the benefits of fire can be replicated mechanically, such as by thinning woody plants or removing invasive species, fire is generally faster and less expensive. Considerations like these led federal land managers to rethink their war on fire in the 1960s, coincident with the ecological turn in federal land management we described earlier in the book. In 1963, for example, the "Leopold Report" on wildlife management in the national parks called on the National Park Service to abandon its approach to wildlife *protection* and instead focus on maintaining and restoring natural ecological processes: "As a primary goal, we would recommend that the biotic associations within each park be maintained, or where necessary recreated, as nearly as possible in the condition that prevailed when the area was first visited by white man."[14] And this included fire: "Of the various methods of manipulating vegetation, the controlled use of fire is the most 'natural' and much the cheapest and easiest to apply."[15] In other words, rather than *protecting* wildlife from wildfire, the Leopold Report recommended that the National Park Service should use fire to sustain the ecosystems on which wildlife depends.

Sure enough, the National Park Service altered its fire policy in 1968 to embrace natural fire within certain limits. Sequoia and Kings Canyon National Parks immediately established a "natural fire management zone" in which they would allow naturally occurring fires to burn unless they threatened to escape the designated area. The Forest Service began experimenting with a similar approach in 1972 and officially abandoned its 10 a.m. policy in 1978. The U.S. Fish and Wildlife Service and the Bureau of Land Management followed suit. The federal agencies had shifted from an official policy of fire *suppression* to one of fire *management,* allowing for the use of fire as a tool.

The new policy faced immediate tests and remains controversial today. In 1978, lightning started the Ouzel Fire in a natural fire zone within Rocky Mountain National Park. Park staff monitored the fire for a month, seeing no significant risks, but then high winds pushed the fire toward the town of Allenspark. The Park Service and other fire services quickly shifted to suppression. Ultimately the town was spared, but as with many wildland fires, this outcome was more likely attributable to weather and topography than to suppression efforts.

Fires in Yellowstone in 1988 also generated enormous public opposition to the "let burn" policy. In June, under drought conditions, lightning ignited fires in the park and in the Forest Service's Absaroka Beartooth Wilderness just north of the park. Based on established management plans, park and forest managers monitored the fires rather than suppressing them, and eleven of the park fires burned themselves out as expected. But fire conditions grew dangerous, ultimately making suppression all but impossible. Thousands of firefighters (up to 9,500 at one time) from federal land agencies, the military, and the national guard dug more than 650 miles of fire line by hand and another 136 miles with bulldozers. Aircraft flew more than 18,000 hours in the park alone, dropping roughly 1.4 million gallons of fire retardant and 10 million gallons of water. Still, firefighters almost lost the iconic Old Faithful Lodge, and fires continued to rage in and around the park.

When September snow finally stopped the fires, almost 1.4 million acres had been impacted. This included 793,880 acres, or 36%, of Yellowstone National Park itself. In the public postmortem, many people blamed the Park Service's "let burn" policy for the fires, even though subsequent research has indicated that, in all likelihood, the Yellowstone fires of 1988 would have been just as bad if the Park Service had maintained a full suppression policy. Still, the episode revealed the political risk that federal agencies face when deciding to allow a naturally occurring fire to burn.[16]

Prescribed natural fire, now generally referred to as "managed fire," remains a significant tool for federal land agencies. Of course, it is not without risk. Federal agencies have a good track record, but it isn't perfect. This was true in the case of Yellowstone, and it is true today. In July 2025, lightning started a small wildfire on the North Rim of the Grand Canyon. Authorities elected to let the fire burn in order to consume excess vegetation and underbrush, but as you can guess by now, winds shifted unexpectedly. In a matter of hours, the fire had shifted course and destroyed the iconic, century-old Grand Canyon Lodge perched on the North Rim, as well as dozens of other cabins and structures. The National Park Service closed the entire North Rim area for the remainder of the season, and, predictably, numerous politicians demanded a complete investigation into the decision to allow the fire to burn.

BOX 10.2 Managed Fire in Yosemite National Park

The Yosemite Fire Management Plan, approved in 2004, divides the park's 750,000 acres into two units: the suppression unit (17%) and the fire use unit (83%). As the name suggests, fires that start in the suppression unit are put out immediately, regardless of ignition source. This unit includes resources popular with tourists, such as the floor of Yosemite Valley, Tuolumne Meadows, and the area around the Wawona Hotel. By

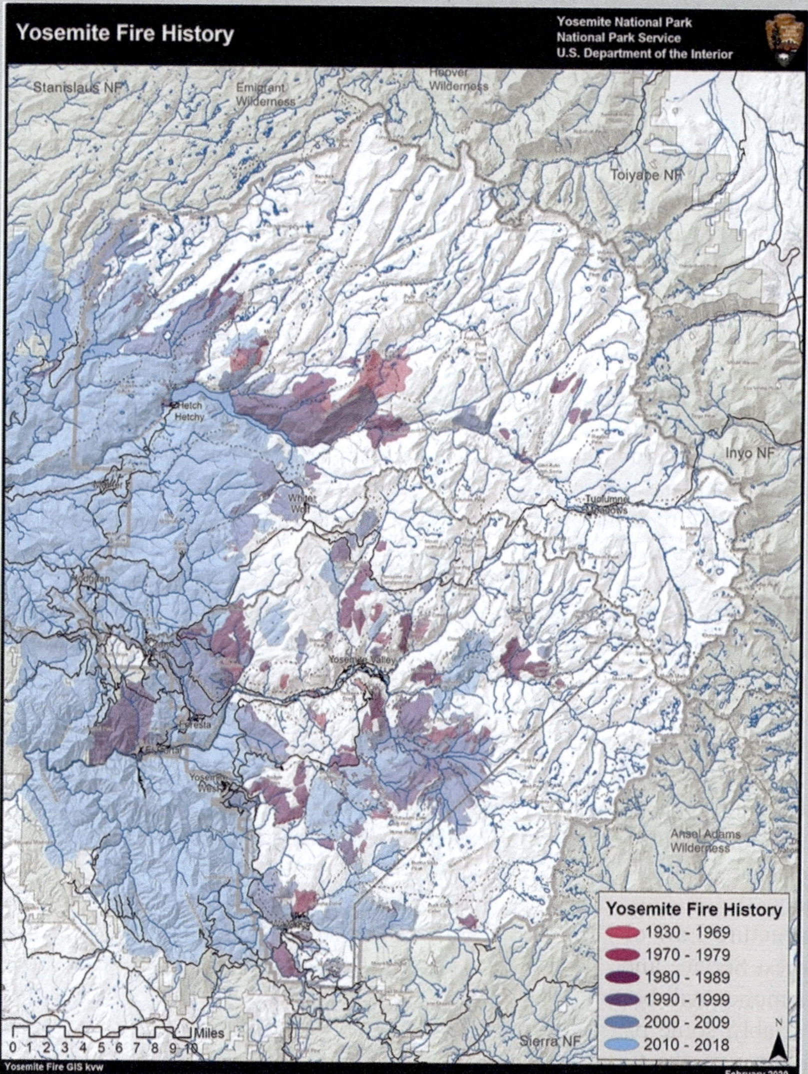

Figure 10.7 Fire history in Yosemite National Park. Source: National Park Service.

contrast, most land in the fire use unit is designated wilderness, which by definition is managed to favor natural processes and conditions. When lightning starts a fire in this unit, trained fire staff must assess the conditions and the risk that the fire could spread outside the unit. Above a certain risk threshold, park staff will initiate suppression efforts. Below that threshold, the staff monitors the fire and allows it to burn under specific parameters.

As Figure 10.7 indicates, wildfire has been widespread within Yosemite in recent decades. Park officials have often allowed natural fire to burn, as it has, for example, in the 40,000-acre Illilouette Creek Basin. Over half the basin has burned in more than two dozen fires across the past four decades, and the results are significant. Research shows that these fires have altered the balance between forest and meadow but have kept fuels from building up, reducing the risk of larger fires in the future.

Prescribed Fire

The more widespread use of fire as a management tool is called "prescribed fire" or "controlled burning." This is when managers plan a fire to accomplish particular objectives and ignite it when conditions are favorable. In the Hart Mountain Antelope Refuge in southern Oregon, for example, the FWS has used prescribed fire to reduce climax vegetation like sagebrush and increase the rich forbs for pronghorn. (No, there aren't any antelope in the Hart Mountain Antelope Refuge. They're really pronghorn.) In Yosemite National Park, managers regularly use prescribed fire on the valley floor to reduce fuels and to open meadow areas. And further south in Sequoia National Park, the Park Service is increasing its use of prescribed fire in and around groves of giant sequoias in an effort to abate the steady loss of these remarkable trees in extreme wildfires.

Playing with fire is dangerous, so federal agencies invest significant time planning each prescribed fire. They analyze the fuels that will burn, the topography, the weather, and many other factors. They consult with air quality regulators to ensure that smoke emissions will comply with the Clean Air Act.[17] Before lighting the fire, they set up containment lines and review contingency plans in case the fire does not behave as expected. They also identify a burn window when weather conditions are likely to remain favorable. The final decision to ignite, the "go/no-go" choice, hinges primarily on the weather. When the active burning is finished, staff ensure that hot spots are extinguished and monitor the area for any signs of reignition.

Federal agencies are increasing their prescribed burning, particularly in the West. The numbers add up. From 2016 to 2019, the four main federal land agencies and the Bureau of Indian Affairs burned an average of 1.7 million acres each year. (Unsurprisingly, the Forest Service carried out the most prescribed fires.) The Bipartisan Infrastructure Law of 2022 allocated additional fire-related funding—$3.5 billion for the Forest Service and $1.5 billion for Department of the Interior agencies—and some of this money will be spent on prescribed fire.

Invariably, some prescribed fires escape their containment. Perhaps the highest profile failure was the Cerro Grande Fire of 2000, where a planned 900-acre prescribed burn in the Bandelier National Monument exploded into a 48,000-acre fire that burned part of the town of Los Alamos, New Mexico, and threatened to destroy the Los Alamos

National Laboratory. In 2022, the Calf Creek and Hermit Peak fires in New Mexico were ignited by prescribed burning operations on the Santa Fe National Forest in New Mexico. They merged to become the largest wildland fire in New Mexico history at roughly 340,000 acres. Congress quickly passed the Hermit's Peak/Calf Canyon Fire Assistance Act, which provided up to $2.5 billion in compensation for those who were injured or lost property.

Such high-profile disasters, as with managed fire disasters, have had an outsized impact on public risk perceptions. In truth, prescribed fire operations have a remarkably good safety record. One study in 2012 found that only 0.8% of the 16,626 prescribed fires conducted by federal agencies escaped. And broader data suggests that the rate of escape remains at or below 1% (a figure which seems small, but to a neighbor of a prescribed burn, it might well appear too high a risk!).[18]

Although federal prescribed burning programs are common, significant obstacles remain. When urgency is required, the fire planning process itself may be an obstacle. Wildland fire suppression has an emergency exemption from NEPA analysis, and Fire Management Plans governing managed fire are approved before those fires start. But choosing to light a prescribed fire is an agency decision that requires NEPA consideration. Some prescribed burning, particularly in the WUI, falls under a categorical exclusion. But many prescribed fires require an EA or even an EIS, which can significantly extend the planning and preparation time. The land management agencies also face challenges where prescribed fires may conflict with air quality standards or other environmental regulations.

Managing Combustibility

There are contexts in which prescribed fire is not a viable option. This is true not only near highly-populated areas, such as gateway towns outside national parks, but also in the WUI, where combustible fuels extend right up to homes and other structures. In some places, smoke from prescribed burns poses an unacceptable health risk. On rangelands, prescribed fire may be rejected because it favors combustible invasive species.

In such instances, managing combustibility becomes essential. On forested land, federal employees or contractors treat areas by physically altering the fuel matrix. In thick stands of trees, crews might use chainsaws to thin the stand and dispose of the woody material, either by stacking it and burning it in the winter or by physically removing it from the forest for use in wood pulp. In some cases, the problem isn't tree density but a dense understory that would allow fire to climb quickly to the forest canopy. Here, the treatment might involve "masticating" the woody material, which is just a fancy term for grinding it up so that it lies flat on the forest floor. In rangelands hard hit by combustible invasives, the agency might choose a chemical treatment such as an herbicide to reduce those particular fuels.

These approaches have their own challenges, especially fiscal ones. Where there is a market for wood pulp or the land manager can offer merchantable timber within the treatment area, agencies may have low-cost options for fuel reduction. And some public land managers have relied extensively on volunteers to collect deadwood fuel in high-traffic areas. But without such cost-saving opportunities, mechanical treatments tend to be expensive because they are so labor-intensive.

BOX 10.3 The Fireshed Registry

Wildfire does not respect ownership or jurisdictional boundaries—it is driven by weather and fuel, not by lines on a map. Once a fire starts, federal, state, and local governments have various tools to coordinate their suppression efforts, including mutual aid agreements and detailed cost-sharing arrangements. Unfortunately, governments have not yet developed such cooperative tools for wildfire *risk mitigation*. For the most part, mitigation planning is something done by particular owners and managers within their own boundaries. This results in serious inefficiencies. For example, fuel reduction work on federal lands may be largely a wasted effort if those lands are adjacent to private property with extreme fire risk.

A new approach, developed largely by the U.S. Forest Service, focuses on coordination within "firesheds." This approach mimics ones used for other environmental problems and natural disasters. For example, agencies delineate watersheds and flood control districts that follow natural boundaries rather than political boundaries, and use these delineations to guide cross-jurisdictional work. In recent years, the Forest Service has developed a "Fireshed Registry" for the continental United States. The registry identifies 7,688 firesheds (each about 250,000 acres), which are spatial units defined by community exposure to wildfire risk. Planners can then examine and classify fireshed attributes, including the risk of wildfire transmission to developed areas, the history of fuel treatments, the community exposure profile, and so forth.[19]

The Fireshed Registry does not guarantee coordination of wildfire risk mitigation, but it has provided an important tool for government agencies at the federal, state, and local level as they plan wildfire risk mitigation work. It has been particularly helpful for federal agencies in prioritizing mitigation projects. For example, the Forest Service has identified the ten firesheds in each forest region that pose the greatest risk to buildings and adjacent communities.

The Challenges Ahead

Wildland fire presents federal land agencies with myriad challenges. The legacy of fire suppression has left some landscapes more vulnerable to catastrophic wildland fires; funding and some elements of agency culture continue to incentivize suppression over a more holistic approach to fire management; climate change is altering fire regimes around the country, often increasing the risk of large, high-intensity fires; and the WUI continues to expand, putting people and property at risk, including wildland firefighters and other first responders. In short, many factors contribute to wildfire risk. It's wise to be skeptical of simplistic characterizations of, and solutions to, the wildfire problem.

To address wildfire more effectively, federal agencies need to continue the ongoing shift from fire *suppression* to fire *management*, and to develop more collaborative, cross-jurisdictional approaches. In 2009, Congress passed the Federal Land Assistance, Management, and Enhancement Act (the FLAME Act), directing federal agencies to develop a comprehensive and cohesive approach to wildfire management. The Departments of Agriculture and Interior, after a 5-year planning process, released their National Cohesive Wildland Fire Management Strategy in 2014. The document's stated vision is "[t]o safely and effectively extinguish fire, when needed; use fire where allowable; manage our natural resources; and as a Nation, live with wildland fire."

We encourage you to read the National Strategy. We won't try to explain it all here. The key for our purposes is to know that it directs agencies to implement more thoroughly the shift from fire suppression to fire management that happened on paper decades ago. The Strategy's highest priorities include continuing fire suppression while doing more work on vegetation and fuels management (prescribed fire, managed fire, and mechanical treatments), reducing human-caused fire ignitions, and providing educational material and incentives to encourage fire-adapted development in the WUI. These objectives extend beyond the federal land agencies' own property, and thus require the cooperation of other federal agencies, state and local governments, and even private entities.

Going forward, the agencies will likely need to realign staffing models to accord with the National Strategy. The legacy staffing model of wildland fire management emphasized seasonal wildland firefighters; as of 2021, only about half of federal wildland firefighters were in permanent, full-time positions. Fully one-third were in temporary, seasonal positions, hired and laid off every year. Today's demands require a different staffing model for several reasons. First, the fire season has gotten longer, so agencies need staff for suppression work for much of the year. Second, since some of the tools of fire management are most effective outside of the main fire season—such as prescribed fire and mechanical fuel reduction—the agencies are likely to need more full-time staff. Special funding authorized by Congress in 2021 and 2022 is equipping agencies to make significant progress. The agencies have committed to converting at least 1,000 positions from seasonal to full-time, and plan to require the people in these positions to spend at least 800 hours each year on reducing hazardous fuels.

Conclusion

The challenges posed by wildfires on federal lands are complex and multifaceted, requiring a shift in approach from mere suppression to a more holistic fire management strategy. The historical success of fire suppression efforts has led, regrettably, to a dangerous accumulation of fuel, making current fires more intense and destructive. This, coupled with expanding human development in the wildland–urban interface and the exacerbating effects of climate change, has elevated wildfire from a nuisance to a major crisis with significant ecological, economic, and social costs. Federal agencies, notably the Forest Service, have recognized the need for adaptive and collaborative solutions, experimenting with managed fires, prescribed burns, and mechanical fuel reduction to mitigate risks and restore ecological balance.

Moving forward, then, effective wildfire management will involve the strategic use of fire and the promotion of fire-adapted communities. Increasingly, agencies are likely to engage in year-round fire management activities with permanent staff. Furthermore, greater collaboration across jurisdictional lines, and the private/public divide, will be crucial for a comprehensive approach to wildfire risk. Only through such integrated and adaptive approaches can federal land managers navigate the "new normal" of intense wildfires and protect both natural resources and human lives.

Study Questions

1 What are the root causes of the wildfire crisis?
2 How should the Forest Service handle the tension between the general public's demand for wildfire suppression and the ecological benefits of wildfire?

3 Describe the managerial problems that confront the Forest Service in light of the growing wildland–urban interface, and explain how Congress might help the Service address these problems.

Resources for Further Study

Confronting the Wildfire Crisis: A Strategy for Protecting Communities and Improving Resilience in America's Forests (Washington, D.C.: U.S. Forest Service, 2022).

The National Strategy: The Final Phase in the Development of the National Cohesive Wildland Fire Management Strategy (Washington, D.C.: The Department of Agriculture and Department of Interior, 2014).

Bradshaw, Karen M. "A Modern Overview of Wildfire Law." *Fordham Environmental Law Review* 21, no. 3 (2010): 445–478.

Egan, Timothy. *The Big Burn: Teddy Roosevelt and the Fire That Saved America* (New York, N.Y.: Houghton Mifflin Harcourt, 2009).

Pyne, Stephen J. *America's Fires: A Historical Context for Policy and Practice* (Durham, N.C.: Forest History Society, 2010).

Pyne, Stephen J. *Tending Fire: Coping with America's Wildland Fires* (Washington, D.C.: Island Press, 2004).

Notes

1 Michael B. Gerrard, "Wildfire Smoke and U.S. Law," *Columbia Journal of Environmental Law* 50 (2025): 209, 222.

2 "Wildfires" (Washington, D.C.: Congressional Budget Office, 2022).

3 Data taken from the National Interagency Coordination Center's wildland fire summary and statistics annual reports. The number of wildfires, averaged over the previous 10 years, dropped from 11,849 in 2011 to 9,598 in 2021. The total number of acres that burned on federal lands, averaged over the previous 10 years, rose from 3.6 million in 2011 to 4.2 million in 2021.

4 "Wildland Fire: Federal Agencies' Efforts to Reduce Wildland Fuels and Lower Risk to Communities and Ecosystems" (Washington, D.C.: Government Accountability Office, 2019), 26.

5 Volker C. Radeloff et al., "Rapid Growth of the U.S. Wildland–Urban Interface Raises Wildfire Risk," *Biological Sciences* 115, no. 13 (2018): 3314.

6 Priya Krishnakumar, "We Mapped Every Wine Country Fire: They're Larger and More Destructive Than Ever," *Los Angeles Times*, November 7, 2019.

7 See generally William M. Downing et al., "Human Ignitions on Private Lands Drive USFS Cross-Boundary Wildfire Transmission and Community Impacts in the Western U.S.," *Scientific Reports* 12, no. 1 (2022); R. Chelsea Nagy et al., "Human-Related Ignitions Increase the Number of Large Wildfires across U.S. Ecoregions," *Fire* 1, no. 1 (2018); Jennifer K. Balch et al., "Human-Started Wildfires Expand the Fire Niche across the United States," *Proceedings of the National Academy of Science* 114, no. 11 (2017).

8 U.S. Office of Management and Budget, "Budget of the United States Government, Fiscal Year 2003," (Executive Office of the President, 2003) 66–67, cited in Karen M. Bradshaw, "A Modern Overview of Wildfire Law," *Fordham Environmental Law Review* 21, no. 3 (2010): 445, 463.

9 For a readable overview of these options, we recommend Stephen J. Pyne, *Tending Fire: Coping with America's Wildland Fires* (Washington, D.C.: Island Press, 2004).

10 The Big Blowup is worth your curiosity, and we recommend both Timothy Egan, *The Big Burn: Teddy Roosevelt and the Fire That Saved America* (New York, N.Y.: Houghton Mifflin Harcourt, 2009), and Stephen J. Pyne, *Year of the Fire: The Story of the Great Fires of 1910* (New York: Viking, 2001).

11 See Joseph B. Ely, "'A Whole New Way to Fight Fire': The Development of Air Tankers in California," *Journal of Forest History* 27, no. 2 (1983).

12 In fact, the system is so effective that it has been adopted by all federal agencies that deal with emergency response, including FEMA.

13 "Wildland Fire: Barriers to Recruitment and Retention of Federal Wildland Firefighters" (Washington, D.C.: Government Accountability Office, 2022).

14 A.S. Leopold et al., "Wildlife Management in the National Parks" (Washington, D.C.: Department of the Interior, 1963), 3.

15 Leopold et al., "Wildlife Management in the National Parks", 8.

16 Jan W. van Wagtendonk, "The History and Evolution of Wildland Fire Use," *Fire Ecology* 3, no. 2 (2007). See also Paul Schullery, "The Fires and Fire Policy," *BioScience* 39, no. 10 (1989).

17 For a detailed discussion of the regulation of wildfire smoke, see Gerrard, "Wildfire Smoke and U.S. Law."

18 John R. Weir et al., "Liability and Prescribed Fire: Perception and Reality," *Rangeland Ecology & Management* 72 (2019): 533–538.

19 Alan A. Ager et al., "Development and Application of the Fireshed Registry" (Fort Collins, CO: U.S. Forest Service, Rocky Mountain Research Station, 2021).

11 Biological Diversity and the Public Lands

Introduction

Imagine Yellowstone National Park without bison, Pelican Island National Wildlife Refuge without white pelicans, or Sequoia National Park without giant sequoias. Millions of people visit public lands each year expecting to see these iconic flora and fauna. Furthermore, a host of species rely on federal public lands for their well-being and even, at times, their survival.

Numerous public lands were set aside by Congress or a president to protect specific species. But federal agencies today are responsible for protecting something more complex: biological diversity (or, more simply, biodiversity), which is the broad variety of life on Earth. Diverse species cannot exist without underlying genetic diversity and the myriad ecosystems that provide each species with critical shelter and nutrients. This is why the first purpose declared in the Endangered Species Act is not protecting species but providing "a means whereby the ecosystems upon which endangered and threatened species depend may be conserved."[1]

This chapter begins with some important background on the rise of biodiversity protection as a federal goal. We then turn to explain the most important statutes directing federal land agencies to protect biodiversity. Finally, the chapter examines some of the key challenges agencies face in carrying out these legal mandates.

The Development of Biodiversity Protection

To understand the vital role that federal public lands play in protecting biodiversity—as well as their considerable limitations—it is helpful to begin with the origins of biodiversity protection as a framework for federal land management. Interestingly, the concept is relatively new; the term *biodiversity* did not gain broad usage until the 1980s.[2] The federal land systems were not designed to address biodiversity protection in any broad or integrated sense. Instead, federal land was regarded as a tool that could be used to protect specific, desirable species.

The Museum/Zoo Model of Wildlife Preservation

In the late nineteenth and early twentieth centuries, wildlife conservation efforts could be divided into two broad categories. The first emphasized production and population. In this view, the goal of wildlife management was to ensure an adequate supply of fish and game for commercial and recreational activities. For example, fish hatcheries

DOI: 10.4324/9781003392514-11

emerged as a way to breed fish for sport and commercial purposes. Similarly, as discussed in Chapter Six, hunters strongly supported the establishment of wildlife refuges to preserve valuable habitat for migratory birds.

But there was another model as well. For some conservationists, the objective was not simply to preserve the abundance of animal populations for human use; the more important goal was preserving species from extinction. The looming possibility that certain species could be eradicated entirely led to support for zoos in which small populations of wildlife could receive direct human care. Museums, too, emerged as places where the public could view exotic wildlife, albeit in taxidermized form. Today, zoos and museums may seem quaint and even backwards as an instrument of conservation policy, but they were influential in elevating public support for the protection of certain forms of wildlife.[3]

BOX 11.1 Bison at the Smithsonian

William Temple Hornaday (1854–1937) was a zoologist and taxidermist who became one of the most fervent pioneers of the American conservation movement. In his role as Chief Taxidermist at the Smithsonian Institution in the 1880s, Hornaday undertook an expedition to the American West to memorialize a species that was thought to be on the precipice of annihilation: the American Bison. The decimation of the once-vast herds—a commercial slaughter that also deprived Native American tribes of a vital food source—converted Hornaday from a hunter into an advocate for wildlife preservation.

Figure 11.1 William Temple Hornaday's bison exhibit in the National Museum. Source: Smithsonian Institution Archives.

Hornaday's famous bison exhibit was unveiled at the National Museum (which would later be called the Smithsonian National Museum of Natural History) in 1887. Unlike previous static displays, the exhibit featured six bison meticulously posed in a dramatic, naturalistic diorama, representing the last vestiges of a wild herd (see Figure 11.1). The presentation served as a powerful moment of public education and transformed the bison into an emblem of conservation urgency. The exhibit not only galvanized public opinion but also laid the groundwork for the formation of the American Bison Society in 1905, which Hornaday would lead. The resulting efforts led to the first successful large-scale reintroduction of animals into a federal game preserve, establishing an important precedent for future wildlife protection initiatives.

Aspects of this museum/zoo model also became evident in public land management, too. The public lands came to be regarded as places where extinction could be staved off by law while visitors watched. In particular, there was broad support for the preservation of certain "charismatic" wildlife species—that is, species recognized and beloved by the public for their majestic qualities like size, power, or speed. President Teddy Roosevelt's first wildlife refuge at Pelican Island in Florida, for example, protected a key stop for the stunning migratory birds that traveled along the Atlantic flyway. Similarly, one of the earliest management goals for Yellowstone National Park was to stop overhunting of the few remaining bison, as well as elk and deer. Park managers started a captive breeding program and grew hay to feed bison in the winter; the acting park superintendent explained that he intended to treat the new bison as any rancher would handle domestic cattle.[4] Right up until 1970, National Park Service staff at Yellowstone would dump food waste onto platforms for bears to eat, while tourists watched from bleachers and listened to presentations by park rangers. Needless to say, these management practices for bison and bears would not occur today.

"Natural" Wildlife Preservation

By the 1960s, both science and society had evolved considerably and now set wildlife preservation on a different, more holistic footing. Environmental historian Thomas Dunlap writes, "Environmentalism emerged as a movement when people applied an ecological perspective to their lives and society, seeing the world as webs of relationships rather than separate things."[5] This perspective profoundly changed the objectives of public land management. If the world was a collection of interconnected webs, then conservation could no longer simply focus on managing individual animal populations. And if nature was self-regulating through intricate relationships, human interference—such as hand-feeding elk or conditioning bears to rely on human waste—was seen as detrimental, distorting the natural balance and behavior of wildlife populations. The shift demanded a move toward a policy of minimal disturbance, allowing natural selection and dynamic ecological processes, like predation and fire, to occur.

BOX 11.2 The Leopold Report

In the case of federal public lands, the shift toward ecological management can be seen in the influential 1962 report by the Interior Department's Special Advisory Board on Wildlife Management, chaired by the renowned conservationist Aldo Starker Leopold, and often referred to simply as the Leopold Report.[6] At one level, the Board retained a sense of the national parks as living museums, recommending that the parks be maintained "as nearly as possible in the condition that prevailed when the area was first visited by white man. A national park should represent a vignette of primitive America." But the report also called for a new approach: the goal was not simply to preserve individual species that existed prior to European contact; the goal was to preserve or restore biotic associations and ecological processes that had sustained bison. A contemporaneous report by the National Academy of Sciences reinforced this view. The purpose of the national parks, the authors insisted, was not to attract tourists and provide them with amenities; "the purpose of the national parks should be the preservation of nature, the maintenance of natural conditions."[7]

The application of ecological science directly challenged the interventionist, zoo-like management practices of the past. The new goal became the preservation of entire biological communities and the complex processes that sustained them. Because many of these processes were not yet well understood, managers often adopted a hands-off, laissez-faire approach as a default. The "natural" was distinguished from the artificial or manmade. The public increasingly appreciated the role of all species—not just the charismatic megafauna—in maintaining a healthy, functioning ecosystem. The focus of land management broadened from mere species preservation to maintaining ecological integrity and the overall health of the land itself.

Biological Diversity and Ecosystem Management

New developments in science supported the Park Service's continued evolution in ecological management, and this included the contemporary language of biological diversity. Ecologists showed that focusing on species alone wasn't wrong, it was just grossly incomplete. Maintaining viable populations of a given species requires an underlying diversity in that species' genetics. Likewise, maintaining healthy populations of diverse species requires protection of diverse ecosystems, or habitats. Biologists coined the phrase "biological diversity" or "biodiversity" to reflect these interdependent scales of diversity. At the same time, a new subfield emerged called conservation biology, which merges scientific research with explicit conservation advocacy, and conservation biologists worked to develop new land management strategies that could protect diversity at the genetic, species, and ecosystem levels.

Conservation biologists emphasized the need for large areas of high-quality habitat managed to protect species diversity and ecological diversity. But they also emphasized the need for connectivity among preserved areas. This encourages genetic diversity, as individuals of a species interact, and it allows for repopulation if a population in one preserve is eliminated by disease, natural disaster, or some other major event. The emphasis on large area conservation and connectivity put new pressure on federal agencies to look beyond the boundaries of their administrative units and to coordinate with surrounding landowners and managers.

In the 1990s, federal agencies embraced a management approach called adaptive ecosystem management. It rests on the assumption that ecological systems are more complex than we can possibly know. As a result, management decisions need to be treated as hypotheses that the agencies test during implementation. Managers expect surprises and plan to adapt as they gather new data. During this period, federal agencies worked on some of the most ambitious, landscape-level plans to date, including the Northwest Forest Plan and the Interior Columbia River Basin Plan. While the George W. Bush Administration moved away from the language of "ecosystem management," top leadership in the Interior Department continued to recognize the need for large, integrated conservation plans.

Threats to Biodiversity

As the twentieth century gave way to the twenty-first, conversations about biodiversity protection came to be dominated by large-scale threats. The primary threats to biodiversity today include habitat loss and fragmentation, invasive species, human population growth, pollution, human overexploitation, and, most notoriously, climate change. Most of these threats are well beyond agency control. The agencies have no power to alter the course of climate change in any meaningful sense, and there are clear limits to how much they can mitigate its impacts (sea-level rise, altered storm patterns, extreme fire behavior, etc.). As a result, biodiversity protection today is necessarily a mix of direct action on public lands, collaboration among federal land agencies, and collaboration with private, state, and tribal landowners. The challenges are immense. And as the next paragraphs will explain, lawmakers have struggled to keep pace with such challenges.

The Law of Biodiversity Protection

There is no single biodiversity protection law governing public lands or federal land agencies. The agencies' organic acts address biodiversity protection, but each one prioritizes it quite differently. The law that most directly addresses biodiversity protection across all federal land systems is the Endangered Species Act (ESA), which was discussed at length in Chapter Four. This statute requires the agencies to protect certain species and their habitats, but it's a challenging mandate for a variety of reasons, not least because many threats to biodiversity cannot be controlled or mitigated within the boundaries of a given park, forest, or refuge. Moreover, biodiversity protection is pursued more vigorously in some presidential administrations than others.

We won't provide a comprehensive account of the complicated legal landscape of biodiversity protection, which varies from agency to agency. Rather, in the upcoming pages we will expand somewhat on our previous discussion of the ESA before explaining how NEPA and various, more specific laws also guide biodiversity protection across the federal public lands.

The Endangered Species Act

The most powerful statute addressing biodiversity protection on public lands is the Endangered Species Act of 1973, and it is difficult to overstate the significance of this law in the context of federal public lands. The ESA's teeth for public land management come in Section 7, which prohibits *every* federal agency from taking *any* action that is "likely to jeopardize the continued existence of any endangered or threatened species or result in the destruction or adverse modification of habitat of such species."[8] The language here is absolute. In the 1978 case *TVA* v. *Hill*, discussed in Chapter Four, the Supreme Court ruled that the ESA required the Tennessee Valley Authority to terminate a nearly-complete dam project on which $80 million had already been spent because completion of the dam would jeopardize a small fish species.

Federal land agencies are not authorized to make this "no jeopardy" determination on their own. The ESA requires them to consult an ostensibly independent, scientific body: the U.S. Fish and Wildlife Service for terrestrial species or NOAA Fisheries for marine species. Such consultations result in a document called a biological opinion (or "BiOp" for short). These analyses consider not only direct threats to individual plants and animals protected by the ESA but also indirect impacts on genetics and ecosystems that could jeopardize a listed species.

Broadly speaking, the ESA has had two major impacts on federal land agencies. First, it has shifted their management priorities. In the Pacific Northwest, for example, Bureau of Land Management and Forest Service timber production plummeted by around 90% in the 1990s when courts ruled that continued federal logging in the region would jeopardize the northern spotted owl. Agency managers had to change their focus rapidly from timber production to habitat protection. Similarly, in 2010, the Fish and Wildlife Service determined that the greater sage grouse merited ESA protection, but it did not have the resources to complete the listing process. Because federal agencies manage 64% of the bird's habitat, listing could have led to a dramatic reduction in federal grazing permits and placed limits on the Department of Defense's exercises in the bird's range. The specter of listing was enough to motivate the BLM and other landowners to protect the sage grouse's habitat. They made enough progress that the Fish and Wildlife Service ruled in 2015 that listing was no longer warranted. Of course, with ongoing drought and wildfire, this could change again.

Second, the ESA has increased cooperation and coordination between federal agencies. Before the Act's passage, federal land agencies often managed adjacent resources in isolation, disregarding the impact their activities had on neighboring jurisdictions. The ESA forced these agencies to recognize that species and ecosystems rarely adhere to administrative boundaries, requiring them to adopt a more holistic, ecosystem-level approach to land management.

Two examples demonstrate this point. The first comes from the Greater Yellowstone Ecosystem, where bears have always been an iconic species. In 1974, just after the passage of the ESA, the Fish and Wildlife Service listed the grizzly bear as threatened in the Lower 48 states. The species had been largely extirpated from its historic range—estimates suggested that only 136 grizzlies remained. In listing the threats to bear populations, the agency pointed to logging and hunting, including in the national forests. The species' recovery required the coordinated efforts of the NPS, USFS, and BLM, along with state and tribal partners. These efforts were formalized through new collaborative bodies, such as the Greater Yellowstone Coordinating Committee and the Interagency Grizzly Bear Study Team, an interdisciplinary group of federal, state, and tribal scientists.

Thanks to the cooperative implementation of a formal ESA recovery plan, the grizzly bear population in the Yellowstone region has rebounded significantly, growing to over a thousand bears and more than tripling its occupied range since listing. In fact, the FWS has attempted to remove federal protections for the grizzly several times, but federal courts vacated these decisions. These legal challenges often revolved around concerns regarding long-term habitat connectivity and the adequacy of post-delisting management plans, some of which fall under state rather than federal control.[9]

The other remarkable example of the ESA driving interagency coordination involves the northern spotted owl. When the owl was listed as threatened in 1990, federal timber harvesting in its range collapsed. Over the ensuing years, the BLM, NPS, USFS, and FWS collaborated on a comprehensive management document called the Northwest Forest Plan. It addressed the needs of the northern spotted owl and a variety of other vulnerable species on 24 million acres of federal lands in the region by designating different zones for various levels of protection. The plan altered land management plans for the BLM and USFS dramatically. Here again, the ESA imposed management parameters on individual agencies, but it also pushed agencies to think across jurisdictional boundaries to achieve common goals.

Notably, the Northwest Forest Plan went beyond the strict requirements of the ESA. In an effort to avoid revising the Plan if additional species were listed under the ESA, the agencies wrote the Plan with the goal of protecting more than one thousand vulnerable species throughout the northern spotted owl's range. The plan was and remains controversial, and the agencies have revised their land use plans considerably since 1994, but elements of the original plan remain in effect today.

The ESA was written before the language of biodiversity gained common usage, and it was not written to preserve biodiversity per se. It focuses on individual species and their habitat. Implemented narrowly, the ESA does not require agencies to prioritize management decisions that protect and restore biodiversity. But as the Northwest Forest Plan illustrates, when the law is applied to protect an indicator species that depends on old growth forest conditions, like the northern spotted owl, its implementation can go well beyond protection of individual species.

The National Environmental Policy Act

The National Environmental Policy Act (NEPA) has played a key role, albeit a largely procedural one, in biodiversity protection on public lands. Unlike the ESA, NEPA does not require agencies to make decisions that protect biodiversity. As explained in Chapter Four, NEPA requires federal agencies to *consider* environmental impacts, including impacts on biodiversity, by preparing environmental impact statements before undertaking any major, federal action with significant impacts on the human environment.

NEPA is important for biodiversity protection in several key respects. First, agencies prepare EISs for a wide range of actions, most of which are not focused on biodiversity protection. Since NEPA requires the agencies to consider *all* environmental impacts, biodiversity concerns are addressed as a matter of course, including concerns related to surrounding federal, state, and private lands. Historically, NEPA rules have also required extensive public participation, from the initial scoping for an EIS to comments on draft and final EISs. This opens the door for citizens and advocacy groups to raise concerns about biodiversity protection regarding a wide range of agency actions. Although the rules that implement NEPA are in constant flux, the statute remains an important trigger for the publication of environmental impacts, including biodiversity impacts. Again, though, the agencies are only required to *identify*—not *reduce*—the environmental impacts of their proposed actions.

Agency- and Site-Specific Law

The ESA and NEPA are the most important laws for biodiversity protection across all federal lands, but there are also a variety of legal requirements that apply to individual agencies. The National Forest Management Act of 1976 (NFMA), for example, directs the Forest Service to "provide for diversity of plant and animal communities based on the suitability and capability of the specific land area in order to meet overall multiple-use objectives."[10] Current Forest Service planning regulations interpret this to mean that land use plans must provide means to "maintain a viable population of each species of conservation concern within the plan area."[11] Notice that the language of NFMA is both more and less stringent than the ESA. On one hand, while the ESA's protection applies only to endangered or threatened species, NFMA mandates biodiversity generally, which the Forest Service interprets to apply to "each species of conservation concern." But on the other hand, NFMA's references to "suitability" and "multiple-use objectives" soften the force of the statute, providing plenty of wiggle room for administrators who wish to prioritize timber sales over biodiversity.

Finally, some federal land designations favor biodiversity protection more than others. National Wildlife Refuges were each created with specific species and habitats in mind, and national parks are managed with an emphasis on natural conditions. But federally designated wilderness is in a league of its own. The Wilderness Act of 1964 defines wilderness as, among other things, "an area of undeveloped Federal land retaining its primeval character and influence, without permanent improvements or human habitation, which is protected and managed so as to preserve its natural conditions."[12] In wilderness management, there is a bias in favor of natural ecological processes. While wilderness restrictions can limit the tools agencies can use to restore and protect biodiversity, they also exclude some of the greatest threats to biodiversity.

Evaluating biodiversity protection across the entire federal estate is challenging, not least because of the wide-ranging ecological conditions and management priorities. The best general measure available is the Protected Areas Database of the United States (PAD-US) maintained by the U.S. Geological Survey. The database categorizes *all* land in the United States from Status 1 to Status 4 according to its protection for biodiversity:

- **Status 1** – Managed for biodiversity with natural disturbance events allowed (for example, wilderness areas, some national parks, some state- or nonprofit-owned nature preserves)
- **Status 2** – Managed for biodiversity with management that may interfere with natural processes (for example, managers may suppress wildfire or flood)
- **Status 3** – Permanent protection, but the land is subject to multiple uses (forestry, farming, intensive recreation, etc.)
- **Status 4** – No known institutional mandates to prevent the conversion of natural habitat

The figures are striking. The database reveals that the federal government manages 96% of Status 1 lands, 67% of Status 2 lands, and 84% of Status 3 lands.

The Challenges of Biodiversity Protection

Biodiversity protection is a global challenge. To date, 196 nations have ratified the United Nations' Convention on Biological Diversity of 1992, which obligates signatories to protect biodiversity at the ecosystem, species, and genetic levels. The United States is notably absent from that list, but as we have seen, there are still a variety of statutes that advance biodiversity protection on public lands. In addition, many states have their own endangered species laws, and a variety of non-governmental organizations, such as land conservancies, protect biodiversity by buying conservation lands or negotiating conservation easements that limit land development.

Nonetheless, efforts to protect biodiversity in the United States face serious headwinds. With respect to public lands, we emphasize three broad challenges to biodiversity protection. The first is competing priorities written into congressional statutes and annual budgets along with competing public demands in land use planning. The second is the enduring mismatch between the pattern of federal land ownership and the lands that are most critical for biodiversity. The third is climate change and its irreversible effects.

Competing Priorities and Public Demands

Both the National Park Service and the U.S. Fish and Wildlife Service are mandated by law to preserve biodiversity, as a component of a broader preservation mission. But these agencies are also mandated to accommodate human use. For the Fish and Wildlife Service, that includes hunting, fishing, and even oil drilling and agriculture. The agencies must constantly negotiate such tensions.

The Bureau of Land Management and the U.S. Forest Service have even greater challenges, since their multiple-use mandates require them to preserve biodiversity *and* produce timber, forage, minerals, and other natural resources. Some of these activities, such as timber harvesting, can be used to advance biodiversity protection goals, but activities like open pit mining destroy ecosystems and are thus at odds with biodiversity.

Furthermore, multiple-use mandates provide considerable discretion to land managers and the presidential administrations to whom they are accountable. Consider, for example, controversy over the Forest Service's "Roadless Rule," which prohibited road construction and timber harvesting on some 60 million acres of national forests and grasslands. Roads threaten biodiversity by fragmenting habitat, introducing invasive species, and facilitating other activities that disturb land. The Roadless Rule was established by the Forest Service at the end of the Clinton Administration in early 2001, but as a mere regulation it was vulnerable to rescission—and indeed, it was rescinded in 2025.[13]

Or consider the Bureau of Land Management's Conservation and Landscape Health Rule (commonly referred to as the Public Lands Rule), which was finalized in 2024. The Rule interpreted the Federal Land Policy and Management Act (FLPMA) to require the BLM

> to build and maintain the resilience of ecosystems on public lands in three primary ways: (1) protecting the most intact, functioning landscapes; (2) restoring degraded habitat and ecosystems; and (3) using science and data as the foundation for management decisions across all plans and programs.

The Rule specifically defined conservation as a "use" on par with other uses authorized by FLPMA. While it was in place, the Rule elevated biodiversity protection within the multiple-use framework—but it too was rescinded in 2025.[14]

Furthermore, the appropriations process limits agencies in important ways. Agencies can only do work that Congress funds, so biodiversity protection can rise or fall with each year's appropriations. The BLM's Public Lands Rule required the agency to use "science and data as the foundation for management decisions across all plans and programs," but Congress does not typically provide vast sums of money to conduct this science or collect monitoring data. Funding is also a chronic problem for the U.S. Fish and Wildlife Service in its work implementing the Endangered Species Act. Not infrequently, the agency determines that listing a given species is "warranted but precluded." This means that the agency has found sufficient evidence to believe that listing is warranted, but it does not have the resources necessary to complete the listing.[15]

Patterns of Federal Land Ownership

The second major set of challenges stems from the patterns of federal land ownership and management explained throughout this book. Public land boundaries were not drawn with biodiversity in mind. (National wildlife refuges come the closest, since they were created to protect specific species.) Instead, they were drawn primarily around notable scenery and valuable natural resources, or in the case of public lands managed by BLM, they were drawn by historical accident.

Ecological boundaries rarely align with rectilinear jurisdictional boundaries. As we saw with the grizzly bear, Yellowstone National Park may have a preservation mandate, but the Park alone is not sufficient to protect the grizzly bear. Success in recovering the grizzly bear in the Greater Yellowstone Ecoregion depends on coordination across national parks, forests, and wildlife refuges. Many wildlife refuges serve migratory birds, providing them with critical resting and nesting habitat. But protecting migratory birds obviously requires habitat and protection over their entire migration routes, which often span multiple countries. Or one could look at Olympic National Park, which occupies the center of the Olympic Peninsula. It was created to protect scenery, including beautiful stands of Sitka spruce. By good fortune, it also serves as critical spawning ground for anadromous fish. But Olympic National Park is an island within the Olympic Peninsula. The only way for adult fish to reach their spawning habitat, and the only way for fry to float back to the ocean, is to pass through areas heavily impacted by logging and other disturbances, where they are at much higher risk.

In short, a major challenge for public lands managers is the mismatch between the lands they manage and the habitat needed for biodiversity protection. Although there is some overlap, particularly in the American West and Alaska, much of the critical land and water for biodiversity protection in the southeast is in state or private ownership. The federal government can still support biodiversity protection there through grants, research, regulations under the Endangered Species Act, and even direct assistance (such as might be provided from the U.S. Forest Service's division of state, private, and tribal forestry). But those efforts depend on cooperative landowners and partners.

Even in areas with a high percentage of federal land ownership, the mandates and priorities of the different federal agencies do not always align. Yellowstone National Park is surrounded by national forests, national wildlife refuges, and public lands managed by the BLM. The Park itself has regulations prohibiting hunting, logging, mining, and most commercial activities beyond visitor services. The wildlife refuges have their own regulations, which emphasize species and habitat protection. But the multiple-use lands managed by the Forest Service and the BLM are open to a wide variety of commercial and recreational activities that can pose a threat to biodiversity.

As an extended example of the mismatch we've described, the next few paragraphs will describe how public land managers in the Pacific Northwest have dealt with the many species of endangered or threatened salmon and steelhead in the Columbia River Basin. At first glance, you might think land management has little to do with fish. Yet the management of riparian public lands in the Basin is important for the survival of these anadromous fish because it affects water quantity and quality, as well as the shoreline habitat that juvenile fish require. At the same time, it is clear that public land management will not be enough to restore the depleted fisheries.

Numerous hydroelectric dams block the Columbia and its tributaries. They impede the migration of spawning fish, reduce juvenile survival, and generally alter food webs (see Box 11.3). But dams are not the only problem in the Columbia River Basin. Commercial salmon harvesting has also put enormous pressure on anadromous fisheries for more than one hundred years by reducing the number of spawning adult salmon each year. Hatcheries were established to offset the impacts of commercial harvests and habitat loss, the idea being that hatcheries could produce fish more efficiently than natural spawning. But hatcheries had a significant and negative impact on wild fish populations by reducing their genetic diversity and increasing the spread of disease. Finally, habitat degradation and loss have been devastating to fish populations. Habitat loss occurs due to the artificial hardening of riverbanks, the creation of still-water reservoirs, the sedimentation of spawning beds, and water pollution. Water quality, in turn, is intrinsically linked to the ecological health of the headwater streams and tributaries, which are largely managed by federal land agencies.

BOX 11.3 Dams and Columbia River Salmon

The Columbia River is one of America's largest and most important waterways, but it is far from a free-flowing river. There are well over one hundred dams in the river's basin, many of which were constructed by the federal government in the twentieth century. These dams supply electricity and have helped irrigate the arid parts of the region, but they have also endangered various salmon and steelhead species.

Floyd Dominy was commissioner of the Bureau of Reclamation during its dam-building heyday of the 1950s and 1960s. In 1994, journalist Marc Reisner asked him if the loss of anadromous fisheries was a worthwhile price to pay for the Bureau's numerous dams. He smiled and said, "I'm sure people can survive without salmon … . I think that there's substitutes for eating salmon. You can eat cake." Dominy's simple calculation was that irrigated agriculture provided a better food supply than salmon.[16]

This view minimizes the profound ecological, economic, and cultural value of Pacific anadromous fish (fish that move from the sea to freshwater streams to spawn). For countless millennia, salmon and steelhead have been foundational to food webs in the Columbia River Basin, providing one of the most critical sources of usable soil nitrogen in the Pacific Northwest. Predators eat them by the millions and spread their precious nitrogen throughout the forests of the Cascade Mountains and the coastal ranges. Culturally, anadromous fish were an essential food source for the Basin's Indigenous peoples, and their spawning runs in the spring and fall also led to important cultural exchanges among the tribes that gathered to fish. Economically, salmon were critical for Indigenous peoples, and throughout the twentieth century, they sustained commercial fisheries and canneries. Dams on the Columbia River and its tributaries have provided hydroelectricity, irrigation, river transportation, and flood control, but the damage they have done has been immeasurable as well.

Ecologists at the Forest Service and BLM recognize that the cold, clean water and complex habitat of headwaters on public lands are indispensable to the freshwater life stages—spawning, incubation, and juvenile rearing—of endangered salmon populations. Their efforts focus on reversing historical degradation from past logging, mining, and grazing practices. Both agencies have implemented policies that discourage or restrict certain land-use activities near perennial and intermittent streams. These policies address timber harvest, road construction, and resource extraction in order to maintain stream shading, which keeps water temperatures low to protect cold-water species, and to stabilize stream banks, which prevents the sediment runoff that can smother incubating fish eggs.

But the work of restoring Columbia River fisheries also takes place beyond the public lands. These efforts have emphasized catch limits, better hatchery operations, and fish passages around dams—in other words, the focus has been on reducing direct human harm to the fisheries. More recently, climate change has emerged as a profound yet slower-moving threat to the Basin's fisheries. Federal scientists have warned that a warming climate will "inevitably lead to extinction unless some ecological, evolutionary, or climatic rescue effect occurs," but the problems posed by climate change involve temporal and spatial scales that are difficult for one state or one country—let alone one public land agency—to address.[17] Regulators can manage water flows by mandating certain flow rates from dams, and they can influence water temperatures by regulating riparian management and stormwater runoff. But regulators can do nothing about precipitation, and climate models predict that the Pacific Northwest will receive more of its precipitation as rain and less as snow. This in turn will alter stream and river flows.

The future of climate change is introducing some new and very difficult questions even for those committed to salmon and steelhead recovery. Is it worth, for example, removing the four tall dams on the lower Snake River to improve access to hundreds of miles of spawning habitat when climate predictions raise concerns that the upper Snake River will be too warm for salmon and steelhead within the next one hundred years?[18] Such questions have no easy answer, and federal land management is just one small piece of an intricate puzzle. At the same time, to ignore the role of public lands in restoring salmon populations would be foolish, especially because public lands management is one of the few tools the federal government has to assist this restoration.

Irreversible Ecological Changes: The Challenge of Climate Change

The challenges we have addressed so far are products of history: federal land ownership patterns and statutory mandates that reflect diverse public interests in public lands. Looking ahead, climate change poses an enormous challenge for biodiversity protection. Land use practices such as open-pit mining, clear cutting, and livestock grazing alter the land surface in specific locations, but climate change is reshaping ecological processes and boundaries in ways that impact the entire public land estate.

Many areas of public land, whether congressionally-designated wildlife refuges or agency-designated "areas of critical environmental concern," were identified as special precisely because of the flora and fauna that exist there, and these organisms depend on relatively stable climatic conditions. Climate change, then, has the power to change the fundamental ecological characteristics of public lands and, in doing so, to eliminate the resources for which those lands were originally protected. Consider the San Pablo Bay National Wildlife Refuge, just northeast of San Francisco. It was created in 1974 to protect and enhance wetlands for migratory birds and other species. Yet this 20,000-acre refuge is projected to lose 95% of its wetlands by 2100, and with it, a host of species. Refuge managers are working on plans to mitigate these impacts. In this instance, the most promising option is simply to buy additional acreage further inland, which would allow wetland conditions to migrate inland as sea level rises. This option, however, will not be available at numerous other federal land units.[19]

Federal land agencies have spent more than a century managing ecological change for a variety of purposes. Some threats to biodiversity, such as human overuse, can be mitigated quite easily in practical terms by restricting human use. This may not be politically popular, but the solution is conceptually clear. Loss of habitat in one location can be addressed by protecting or restoring that habitat in another location. Again, this might not be popular politically, but it is conceptually clear. Climate change is very different, since the agencies can do almost nothing to slow climate change in the near-term. In many cases, agencies do not have the tools necessary to stop comprehensive ecological changes that alter the ecological functions and species composition of an area.

As discussed in Chapter Nine, the Resist-Accept-Direct (RAD) framework encourages federal agencies to consider whether climate impacts should be resisted, accepted, or directed toward a desired outcome. Pelican Island, in Florida, was designated as the first federal wildlife refuge by President Teddy Roosevelt in 1903. At the time, it was just over 5 acres in size. Over the next century, though, boat wake, severe storms, and sea-level rise eroded almost 3 acres of the original island and threatened to eliminate it altogether (see Figure 11.2). The FWS has chosen to resist this change with significant intervention. It brought in hundreds of tons of oyster shells to create a storm break that would absorb wave energy and reduce erosion, and it planted vegetation that would accelerate the accretion of sediment on the island's perimeter. Today, the island is about 3.3 acres and growing. If Pelican Island is to continue to serve as a habitat for migrating birds, this type of work will need to be expanded in the next half-century to keep pace with ongoing sea-level rise and higher storm surges.

Where resistance is impossible and acceptance is untenable, federal agencies are innovating rapidly to direct ecological changes in ways that advance their biodiversity mandates. The Forest Service, for example, has historically "planted local." When the agency needed to reforest an area, it used local genotypes of trees because they were best adapted to that location. But trees can live for hundreds of years, and the Forest Service recognizes that the genotypes best suited for today may not survive the climatic changes of the coming decades. As a result, the Forest Service now sometimes uses genotypes from another location, typically to the south of the targeted area.

Similarly, managers at Glacier National Park have accepted that they cannot alter glacial loss, but they haven't accepted the secondary impacts of melting glaciers on the

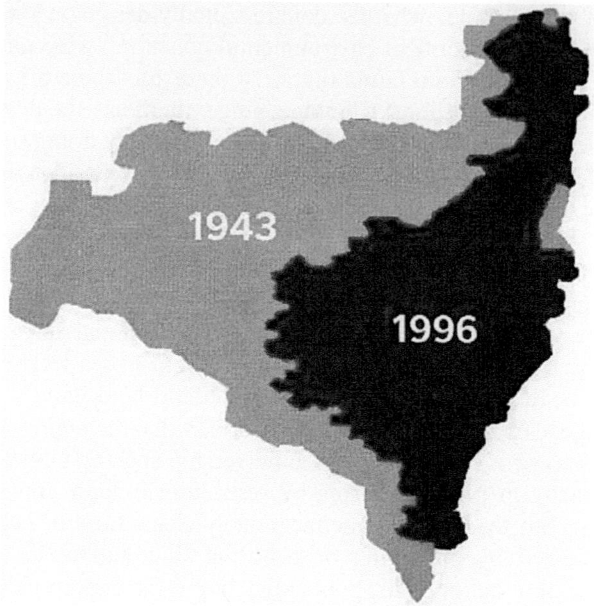

Figure 11.2 The erosion of Pelican Island. The image compares the size of the Island in 1943 and 1996. Source: U.S. Fish and Wildlife Service.[20]

key species within the park. For example, glacial loss is leading to warmer water temperatures in the Park's streams; projections indicate that some streams will become too warm to support a species of bull trout listed as threatened under the ESA. Park managers have decided to move a population of bull trout to a different, higher stretch of its usual stream—a stretch otherwise cut off from the lower creek by a waterfall—in order to provide cooler water. This is no small decision, and it comes with risks, but park managers decided to direct ecological change rather than allow the bull trout to be extirpated from the park.

BOX 11.4 Biodiversity and Climate Change: The Southern White-Tailed Ptarmigan

The southern white-tailed ptarmigan lives principally in the Rocky Mountains of Colorado (see Figure 11.3). Nonmigratory, it lives its entire life on mountaintops and is adapted to harsh conditions that exist above the tree line. It thrives in relatively cold conditions, using snow for shelter, and it experiences physiological stress in temperatures above about 70° F. Its plumage changes seasonally so it is camouflaged against changing ground conditions. It depends on snowpack to produce spring and summer moisture for its forage, and the timing of that forage growth is critical for rearing its young.

This ptarmigan is an excellent example of a species threatened principally by climate change.[21] Most of its land in national parks and forests is in designated wilderness areas, so many of the common human disturbances threatening other species—mining, hunting, grazing, predation, infrastructure development—do not affect the species' viability. Instead, climate change is the most significant threat for several reasons. First, rising temperatures and decreasing precipitation in the Rocky

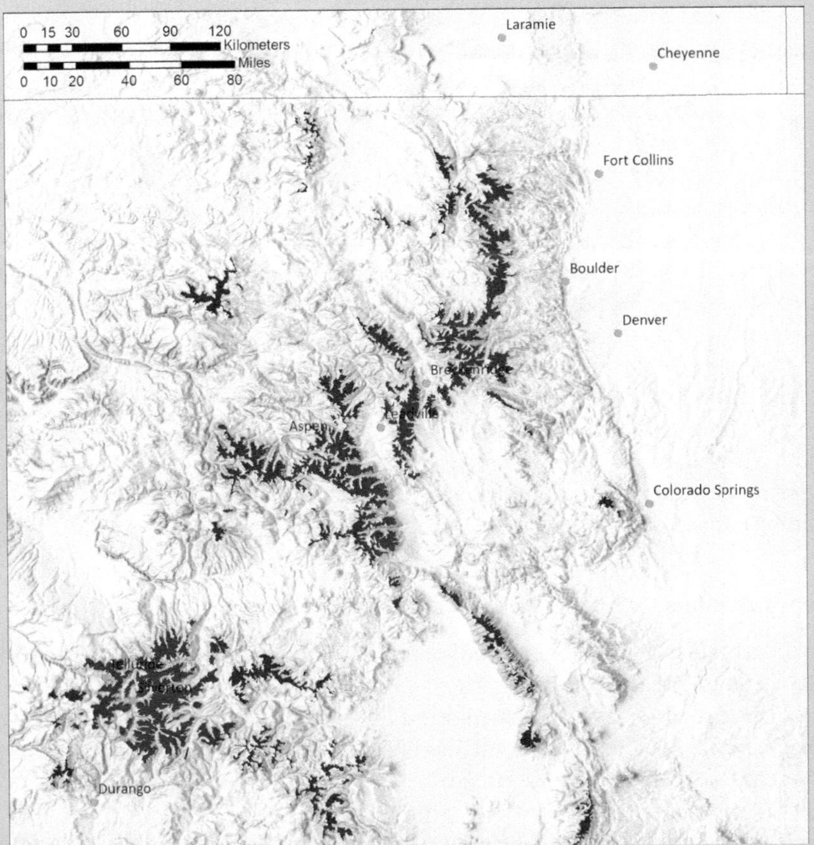

Figure 11.3 The blackened mountaintop areas indicate the range of the southern white-tailed ptarmigan. Source: U.S. Fish and Wildlife Service.

Mountains are reducing snowpack throughout the year, limiting the bird's shelter. Second, snowmelt is peaking earlier in the year. This means that soils dry out earlier, reducing the availability of food sources and diminishing moisture in the late summer and early fall. Third, warmer temperatures cause the tree line to rise throughout the bird's habitat, reducing the open, alpine vegetation it needs.

The Endangered Species Act requires the FWS to develop recovery plans that will improve the viability of threatened or endangered species. But for high alpine species, the recovery tools available to federal land managers are limited. Agencies cannot create new ptarmigan habitat at higher elevations, since the bird already inhabits the highest peaks in the continental United States.

Conclusion

Biodiversity protection is a major challenge for federal public land managers for a host of reasons. The science of biodiversity is dynamic, and ecological systems are even more dynamic. Public demands are diverse and contradictory. The agencies are working with old patterns of land ownership under legal and regulatory structures that address biodiversity in a piecemeal, non-comprehensive fashion. Each presidential administration, and each successive Congress, prioritizes biodiversity protection differently, so administrative goals and their funding swing back and forth over time. Climate change is further complicating biodiversity protection and forcing agencies to make difficult choices about whether to resist, accept, or direct change.

Still, the agencies do have legal mandates to protect biodiversity in certain ways, and they have enjoyed some successes, such as the protection of grizzly bears in Yellowstone National Park. The future of biodiversity protection on federal lands will depend on the adaptive capacity of the agencies, but even more on a renewed societal commitment to ecological health. As climate change continues to dismantle historic ecological stability, preserving biodiversity will require sustained funding for ecological research, a consistent prioritization of conservation as a legitimate and essential "use" of public lands, and innovative, whole-ecosystem collaboration that transcends administrative and political boundaries. The American public lands remain the single most important reservoir of the nation's biodiversity, and their successful stewardship demands a corresponding elevation of biodiversity protection as a national imperative.

Study Questions

1. What are the fundamental limitations of the Endangered Species Act in addressing the various threats to biodiversity?
2. To what extent does the fragmentation of authority among the four major federal land agencies inherently prevent the effective, landscape-scale biodiversity protection called for by ecological science?
3. Considering that the majority of critical habitat exists outside federal boundaries, how should public land policy adapt to better incentivize private conservation and biodiversity stewardship on adjacent lands?

Resources for Further Study

Baier, Lowell E. *The Codex of the Endangered Species Act: The First Fifty Years* (Lanham, MD: Rowman & Littlefield, 2023).

Dunlap, Thomas R. *Saving America's Wildlife: Ecology and the American Mind, 1850–1990* (Princeton, NJ: Princeton University Press, 1988).

Farnham, Timothy J. *Saving Nature's Legacy: Origins of the Idea of Biological Diversity* (New Haven, CT: Yale University Press, 2007).

Hilty, Jody A., Charles C. Chester, and Molly S. Cross (eds.). *Climate and Conservation: Landscape and Seascape Science, Planning, and Action* (Washington, D.C.: Island Press, 2012).

Keiter, Robert B. *Conserving Nature in Greater Yellowstone: Controversy and Change in an Iconic Ecosystem* (Chicago: University of Chicago Press, 2025).

Skillen, James R. *Federal Ecosystem Management: Its Rise, Fall, and Afterlife* (Lawrence, KS: University Press of Kansas, 2015).

Notes

1 Endangered Species Act of 1973 § 2(b).
2 Sahotra Sarkar, "Letters: Origin of the Term Biodiversity," *BioScience* 71 (2021): 893.

3 Ben A. Minteer, Jane Maienschein, and James P. Collins, *The Ark and Beyond: The Evolution of Zoo and Aquarium Conservation* (Chicago, IL: University of Chicago Press, 2018).
4 Mary Ann Franke, "Captivity for Yellowstone Bison: Their Doom or Salvation," *Montana: The Magazine of Western History* 55, no. 2 (2005).
5 Thomas R. Dunlap, *Saving America's Wildlife* (Princeton: Princeton University Press, 1988), 95.
6 A. Starker Leopold et al., "Wildlife Management in the National Parks," reprinted in Lary M. Dilsaver, ed., *America's National Park System: The Critical Documents* (Lanham, MD: Rowman & Littlefield, 1997).
7 National Academy of Sciences, *A Report by the Advisory Committee to the National Park Service on Research* (Washington, D.C.: National Research Council, 1963). This report is referred to as the "Robbins Report" after its chairman, William Robbins.
8 Endangered Species Act of 1973 § 7(a)(2).
9 See generally Robert K. Keiter, *Conserving Nature in Greater Yellowstone: Controversy and Change in an Iconic Ecosystem* (Chicago: University of Chicago Press, 2025).
10 16 U.S.C. § 1604(g)(3)(B).
11 36 C.F.R. § 219.9(b)(1).
12 Wilderness Act of 1964 § 2(c).
13 The initial Roadless Rule can be found at 66 Fed. Reg. 3244 (2001). Its rescission is at 90 Fed. Reg. 42179 (2025).
14 The Conservation and Landscape Health Rule can be found at 89 Fed. Reg. 40308 (2024). Its rescission is at 90 Fed. Reg. 43990 (2025).
15 See Kristina Alexander, *Warranted but Precluded: What That Means Under the Endangered Species Act* (Washington, D.C.: Congressional Research Service, 2010).
16 The interview with Dominy aired on the 1997 PBS documentary, "Cadillac Desert," which you can find on YouTube.
17 Lisa G. Crozier et al., "Climate change threatens Chinook salmon throughout their life cycle," *Communications Biology* 4 (2021): 222.
18 The possibility of breaching the lower four Snake River dams has provoked fierce controversy in Washington and Idaho for decades. See, e.g., Hannah L. Hilbert-Wolf and Andrea K. Gerlak, "The evolution of the modern dam conflict on the Snake River, USA," *Water International* 47, no. 8 (2022): 1349–1369.
19 For additional context, see Karen M. Thorne et al., "Are coastal managers ready for climate change? A case study from estuaries along the Pacific coast of the United States," *Ocean and Coastal Management* 143 (2017): 38–50.
20 U.S. Fish and Wildlife Service, *Pelican Island National Wildlife Refuge: Comprehensive Conservation Plan* (Washington, D.C.: U.S. Dept. of Interior, 2006), 13.
21 As of this writing, the southern white-tailed ptarmigan is not listed, but its cousin, the Mount Rainier white-tailed ptarmigan, is listed as threatened. As the name suggests, this subspecies' habitat is primarily in the alpine regions of the Pacific Northwest.

12 Recreation on the Public Lands

Introduction

The public lands serve as the nation's playground. You can ski, hike, or camp; hunt, snowmobile, or rock climb; ride an ATV, canoe, or airplane. You can kayak through whitewater rapids, bike down a mountain, or simply gawk at wildlife or scenic vistas. The possibilities—and the destinations—are seemingly endless. But what forms of recreation should be allowed on the public lands, and in what amounts? What restrictions are appropriate? Should some activities be prioritized over others? Should land managers allow commercial enterprises to operate on public lands, and if so, under what conditions? These questions have no easy answers. And the questions themselves are always evolving, given Americans' ever-changing recreational interests.

This chapter explores recreational policy on the federal public lands. It begins with an overview and then turns to describe how agencies make policies about recreational use. The rest of the chapter then focuses on three general issues implicated by recreation. The first is overcrowding, which is increasingly a fact of life on many public lands. Second, and relatedly, is the issue of access to recreational opportunity—an issue that implicates not only concerns about equity but also the practicalities of transportation to, from, and within the public lands. Finally, the chapter will discuss the many private concessioners that operate on the public lands and the policies that affect these businesses.

The Dynamic Nature of Public Land Recreation

It was not foreordained that, in the public mind, recreation would become the predominant use of the public lands. In the writings of the great American naturalists of the twentieth century—people like John Muir, Edward Abbey, Rachel Carson, Aldo Leopold—one can find various aspirations and hopes for the public lands, and various organizing principles from which management decisions might proceed. Some writers emphasize the spiritual aspects of exposure to nature, seeing in the public lands an opportunity for wild lands to teach visitors something about the world and their relationship to it. Others conceive of human visitation to pristine places as something more like trespass, an undesirable intrusion into an almost sacred space. Whatever one might say about such philosophies, the raw fact is that hundreds of millions of people visit the public lands each year, and their uses are overwhelmingly recreational. Massive recreational use of the public lands is now simply a reality—a social fact to be managed and shaped rather than undone.

DOI: 10.4324/9781003392514-12

In the early days of the National Park Service, its inaugural director, Stephen Mather, used marketing campaigns to highlight the parks' recreational appeal. One of his goals was to expand political support for the parks by encouraging recreational visitation. Then as now, policies about recreation on all the public lands—not just the national parks—are tethered to the democratic process. Management decisions are not dictated directly by popular opinion, of course, but philosophical choices about the purposes of public land management are ultimately driven by elected politicians and their appointees.

Although recreation has long been one of the most important categories of public land use, recreational preferences are always changing. The amount of recreational visitation is growing as tourism increases and as more people spend more time outdoors. Particular sites wax and wane in popularity. And the mix of recreational activities is evolving. Sightseeing and hiking have lasting appeal, but other activities come and go as fads change and technologies adapt.

The National Park Service reports that annual recreational visits to NPS properties have grown from about 220 million in 1980 to 325 million in 2023. The Forest Service recently tallied about 170 million annual recreational visits, up from about 148 million just 10 years before. The BLM estimated nearly 80 million visitor days on its public lands in 2022, up from 68 million in 2002. And nearly 60 million people make recreational visits to National Wildlife Refuge system lands each year.

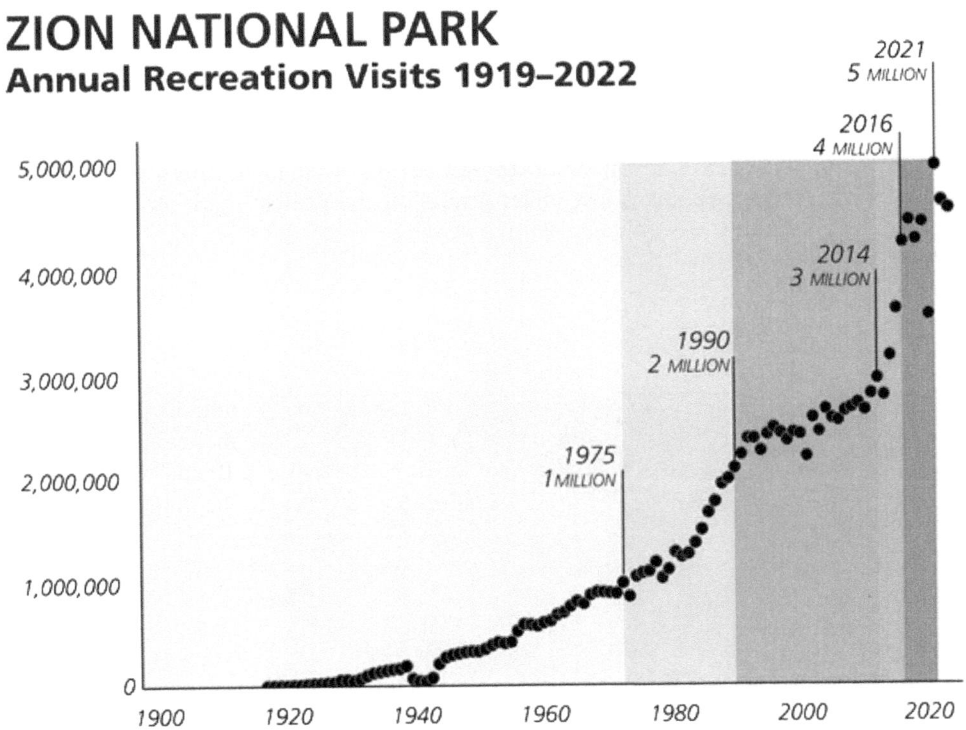

Figure 12.1 Recreational visits to Zion National Park have exploded in recent years, reflecting national trends on all the public lands. Source: NPS.

BOX 12.1 William Tweed on Outdoor versus Online Recreation

William Tweed was the Chief Naturalist at Sequoia and Kings Canyon National Parks until his retirement in 2006. A prolific author, Tweed frequently opined on the public lands and their challenges and prospects. His 2010 book, *Uncertain Path: A Search for the Future of National Parks*, contains the following passage:

> Current trends suggest that the recreational role of national parks is likely to continue to shrink. In California, the parks ... are more difficult to access than several decades ago because of population growth and traffic congestion. At the same time, entire new competing recreational worlds have blossomed. The virtual reality of the digital world offers much that attracts. One can be transported to another realm without having to travel or even sweat. The rewards of engagement are immediate; why go instead to a faraway place that requires a lot of time and special skills to enjoy?[1]

How wrong he was! Tweed feared that the online world would diminish interest in the public lands, but he underestimated the growth in overall demand for recreation. In hindsight, it seems there is increasing demand for *both* online and outdoor recreation. The decline in park visits mentioned by Tweed was a blip which disappeared almost immediately. Recreational visits to the public lands have grown steadily over the past several decades.

These trends are eye-opening, but they mask even more rapid changes in visitation to particular sites. Sometimes a specific place becomes wildly popular almost overnight. Horseshoe Bend, for example, is an especially picturesque viewpoint along the Colorado River in the Glen Canyon National Recreation Area in northern Arizona (see Figure 12.2). The viewpoint is just under a mile off the nearest road, and it had long

Figure 12.2 Horseshoe Bend (Arizona). Source: Wikimedia Commons (converted to grayscale).

been a sleepy, off-the-beaten-path spot known mostly to locals and boutique photographers. But when social media exploded in the early 2000s, Horseshoe Bend went viral. Visitors thronged in massive numbers to take selfies or to post their sunset shots on Instagram. Summer sunsets drew thousands of visitors—visitors who clogged the two-lane road, parked illegally, walked off the designated paths, and left trash scattered about. The nearby town of Page spent millions to widen the road and provide parking; the National Park Service added guardrails and interpretive content. Sudden changes like these, unimaginable before the internet, have become almost commonplace.

Visitors do much more on public lands than just take pictures, though. Many wish to use motorized vehicles to explore scenic terrain. From four-wheelers to e-bikes, these vehicles vary in the infrastructure they require, the noise and intrusion they represent, and the scars they inflict on the land. Drivers approaching Moab, Utah, are bombarded by billboards advertising jeep trips deep into the red rock country in and around Canyonlands National Park. Snowmobiling in the public lands of northern Minnesota attracts tens of thousands each year. Unsurprisingly, these activities are frowned upon by numerous other visitors—visitors who trekked to Utah or Minnesota precisely to avoid the noise and frenzy associated with motor vehicles.

The rapidity with which recreational preferences evolve requires public land managers to revisit usage policies frequently. And as you now know, policy changes are not easy or quick. Managers may be required to consult numerous stakeholders or to complete a lengthy environmental analysis before revising a single protocol. All the while, public pressures continue to change. In light of such constant evolution, let's examine the process by which recreational policies are made.

How Are Recreational Policies Made?

Unlike biodiversity, energy, or some of the other issues covered in these chapters, there is not a legal framework dedicated to recreational policy. Agency personnel are largely free to make decisions on the basis of their general authority to manage the lands in the public interest. That general authority is broad but remains constrained by all the laws you've been learning about in this book. For example, the Wilderness Act forbids commercial activities, motorized equipment, and mechanical transport within wilderness areas. This means that managers of wilderness can't allow concessioners to offer guided experiences there, nor can they permit bicycles on wilderness trails (see Box 12.2).

BOX 12.2 Mountain Bikes in the Wilderness?

Since its emergence in the 1970s and 80s, mountain biking has become one of America's most popular modes of outdoor recreation. Enthusiasts love to seek out remote, mountainous terrain. But because many mountain bikers enjoy riding on hiking trails, there are occasional conflicts between hikers and bikers. These conflicts occur not only on the trails but also in policy conversations. One particularly active area of dispute involves the question of whether mountain biking should be allowed in federal wilderness areas.

To answer that question, we must examine Section 4(c) of the Wilderness Act. This important section contains the key prohibitions that define federal wilderness:

> Except as specifically provided for in this chapter, and subject to existing private rights, there shall be no commercial enterprise and no permanent road within any wilderness area designated by this chapter and, except as necessary to meet minimum requirements for the administration of the area for the purpose of this chapter (including measures required in emergencies involving the health and safety of persons within the area), there shall be no temporary road, no use of motor vehicles, motorized equipment or motorboats, no landing of aircraft, no other form of mechanical transport, and no structure or installation within any such area.[2]

Although the Act doesn't mention bicycles, the words "other form of mechanical transport" could include them, as bicycles are clearly a mechanical mode of transportation. But bicycles are quite different from motor vehicles, motorboats, and aircraft, which are specifically mentioned, and mountain bikes didn't even exist when the Wilderness Act was passed in 1964. So what should land managers do?

The Forest Service was the first agency to tackle the question directly. In the 1970s, the Service issued a regulation making clear that it regarded bicycles as a form of "mechanical transport" banned by the Act. A few years later, the other land management agencies followed suit and imposed bans of their own. By the mid-1980s, mountain biking was prohibited in most federal wilderness areas around the country.

But the battle didn't end there. Countless mountain biking advocates regard their sport as wholly consistent with wilderness preservation and enjoyment. They fought tirelessly to open wilderness to bikes, and when it became clear that the agencies had no intention of revising their regulations, these advocates went directly to Congress. Several bills were proposed in the late 2010s and early 2020s, each trying to amend the Act to specify that bicycles would be allowed. None of these bills have passed, however, so for the moment, the ban on mountain biking in wilderness continues.

For those myriad instances in which the law does *not* prohibit a particular activity, managers must exercise their ordinary managerial discretion to make policies about recreational activities. What does this look like, and how do agencies reach their decisions?

The decision begins with the general management standard applicable to the particular agency and land unit. For the BLM and Forest Service, that standard is the multiple-use standard; for national parks and wildlife refuges, that standard is instead a dominant-use standard.[3] Recall also that the agencies will likely have engaged in a planning process, yielding a management plan or a similar governing document. The management plan represents the agency's attempt to articulate, in a general way, how it will implement the relevant legal standard in this particular area or land unit. The standard and the plan together represent the starting point for agencies as they consider decisions about any land use, including recreational uses.

Management plans are not self-implementing—that is, they represent only a *plan* for action, not the action itself—but in effect they do rule in and rule out certain kinds of land uses. For example, if a plan specifies a certain area for a timber harvest or mining development, that use will obviously preclude certain recreational activities in that area. Similarly, if an area is not set aside for a particular use, it will generally remain open to the public for a wide array of dispersed recreational activities. As an example, take a look at Box 12.3, which provides an excerpt from a BLM management plan in

Wyoming. Notice how permissive the plan is with respect to general, low-intensity recreational activities, from camping to horseback riding.

BOX 12.3 Recreational Uses in a BLM Resource Management Plan

The following excerpt is taken from a Resource Management Plan revision issued by the BLM's Rock Springs Field Office (RSFO) in Wyoming:

3.17.1 Recreational Use

Types of recreational use include dispersed recreation and developed recreation. Dispersed recreation consists of activities of an unstructured type that are not confined to specific locations or dependent on developed recreation sites. Dispersed recreation occurs throughout the planning area over a wide range of ecosystem types, and includes sight-seeing, touring, backpacking, horseback riding, geocaching, hiking, off-highway vehicle (OHV) use, photography, wildlife viewing, fishing, other water related activities, hunting, and camping. The RSFO manages many developed recreation sites scattered throughout the RSFO, consisting of day use/picnic areas, campgrounds, interpretive sites, and historic site tourism. Developed recreation sites provide excellent opportunities and starting points for activities such as camping, hiking, backpacking, horseback riding, wildlife viewing, sightseeing, OHV touring, fishing, and hunting.

Special recreational permits (SRP) are issued to manage visitor use, protect natural and cultural resources, and achieve the goals and objectives of the recreation. The five general categories of SRPs are commercial, competitive, vending, individual or group use in special areas, and organized group activity and event use.

The planning area administers approximately 20 SRPs per year and this number has remained relatively stable from year to year. Within the planning area SRPs are administered for activities and events such as outfitting and guiding for hunting activities, fishing, floating, horseback rides, wild horse viewing tours, interpretive tours, livestock drives, horseback fund raising events, horse endurance rides, yoga trips, and llama treks … [4]

But what about more intensive activities, locations in very high demand, or other situations in which more specific analysis is required? And what if demand is growing for a recreational activity that is inconsistent with the current management plan? In those instances, agencies must engage in a much more thorough decision-making process. A fresh revision of the management plan may be required, which is a time-consuming and expensive process. If the potential impacts of a decision are sizeable, the agency may have to conduct an environmental assessment to satisfy NEPA, or perhaps an interagency consultation to satisfy the Endangered Species Act. But even if these burdensome processes are not required, the agency will likely consult stakeholders and affected parties and will perhaps invite comments from the general public. There may well be a second round of consultation after the agency develops an initial proposal. These rounds of consultation may not formally be required by law, but they help the agency understand and evaluate the likely reactions to possible courses of action. If the agency engages in a thoughtful and careful analysis, its decision is more likely to withstand any backlash or opposition that may arise, even if that opposition takes the form of a lawsuit. If there is no specific prohibition or mandate regarding a particular decision, the law is on the side of an agency that makes careful, reasoned decisions.

Let's put this all together with a specific example: rock climbing within Yosemite National Park. For many climbers, Yosemite is the best climbing destination in the world. Ascending the face of El Capitan is a bucket-list activity, one that requires enormous skill, specialized equipment, advance planning, and perhaps guided supervision. (The National Geographic documentary *Free Solo* can give you a great sense of what's involved!) Climbing can be done safely, but it is inherently dangerous; Yosemite averages about 2.5 climbing deaths per year. And as climbing has grown in popularity, so have its impacts within Yosemite. Some of those impacts are small: for example, climbers commonly leave bolts in place along climbing routes for use by other climbers. But in recent years it has become commonplace to find more substantial detritus, such as excess rope dangling from a cliff face or scattered waste at the base of a popular route.

So how do Yosemite officials deal with rock climbing? The first task is to see if the law already provides an answer. There is no law that speaks expressly to rock climbing at Yosemite, but many of Yosemite's most popular climbs, including El Capitan, are within federal wilderness areas. As you've seen, the Wilderness Act bans "structures or installations" within wilderness, except when they are "necessary to meet minimum requirements for the administration of the area." Prompted by wilderness advocates in the early 2020s, the National Park Service took up the question of whether climbing bolts should be considered "installations" within the meaning of the Act. In late 2023, the NPS circulated a draft report for public comment. The report suggested that bolts *are* installations, a position the Service had never previously articulated. But the report also contained an application process by which climbers could nominate sites for permanent bolts and argue for their necessity to meet the area's "minimum requirements."

Figure 12.3 El Capitan, Yosemite National Park. With a 3,000-foot granite face, it is one of the most famous climbing destinations in the world. Source: Arun Kuchibhotla.

The draft report pleased no one—not wilderness advocates, who tend toward literalism in interpreting the Wilderness Act and thus felt no exceptions should be made; and certainly not climbers, who not only claim that bolts are not "installations" but also point out that they've been using bolts on wilderness rock for decades with no complaint from the NPS.[5]

While agency heads in Washington, D.C. decide how to handle this issue of national policy, rangers in Yosemite must still have an interim policy on rock climbing within their park. Without a clear legal mandate or a national rule promulgated by the Service, the park superintendent is free to decide how to manage this activity pursuant to the NPS's general management authority. At one extreme, Yosemite staff could ban rock climbing anywhere in Yosemite. No doubt this would be a dramatic and highly unpopular move within climbing circles, but the Park Service could perhaps justify it on the basis of the danger associated with climbing. After all, Yosemite bans BASE jumping anywhere within the park—like climbing, it's an activity with a minimal environmental footprint but one that is dangerous compared to low-intensity recreational activities. Conversely, Yosemite officials could impose no restrictions at all, allowing a free-for-all at El Capitan and other climbing routes.

Here is how the superintendent has chosen to proceed. Yosemite has not banned climbing, recognizing it as a popular sport with widespread appeal. Instead, the Park Service instituted a self-registration permit system to address the minor problems that have arisen. This minimalist scheme requires overnight climbers to complete a registration form near the base of El Capitan. The form's purpose, aside from helping rangers to gather data on climbing activities, is simply to make climbers aware of the concerns associated with increased climbing activity. The permit creates a "touch point," which enables the NPS to educate climbers about their impacts. Just as rangers provide instructions to backpackers entering the backcountry, the form instructs climbers about how to handle excess rope and how to dispose of human waste while on overnight climbs. In sum, park officials have created a tidy, user-friendly management regime that emphasizes education rather than restriction. To make this policy, the agency sought feedback from the climbing community. The self-permit system leverages the strong ethic among climbers, which is often enforced internally and informally, to accomplish the NPS's goals. Meanwhile, the agency continues to seek feedback and intends to refine its policy from time to time to accommodate climbers while also attending to the interests of other park visitors and the conservation of park resources.

Overcrowding

We shift our attention now to one of the most difficult problems confronted by federal land managers: overcrowding. "Loving them to death" is a phrase often invoked to describe the impacts of recreational visitors on popular public lands. To visit Yellowstone, Glacier, Yosemite, Zion, or Grand Canyon National Park during the height of summer is to encounter thousands upon thousands of other tourists, all squeezing in for the same photos, clamoring for the same parking spots, walking down the same trails, and queueing for the same bathrooms. Overcrowding not only imperils park resources; it creates managerial and infrastructural challenges and degrades the visitor experience. And it doesn't occur only at big national parks. Spots like Horseshoe Bend (mentioned above) are vulnerable, too, as is any hot spring, fishing hole, or climbing route that goes viral on the internet.

What's a conscientious land manager to do? On one hand, public visitation is a good thing. Public lands law anticipates and prioritizes human enjoyment. Those who visit the public lands reap personal rewards in the form of rejuvenation, enjoyment, and relaxation, and tend to act as better stewards of natural resources after their visits. But allowing too many people onto the public lands can create risks and bring negative consequences. So managers of popular sites must make some difficult decisions.

For starters, managers must address the ways that crowding affects basic site upkeep. Large crowds mean that public land infrastructure is more likely to be strained, to wear out, to need additional maintenance, or to face intentional harm. When sites become crowded, land managers have to think carefully about maintaining pathways, adding guard rails at viewpoints, increasing trash collection, and so forth. In some instances, increased visitation may require costlier improvements—a larger parking lot, a wider road, or more restrooms. Avoiding such expansion can be an indirect way of discouraging crowds—if a parking lot is small, visitors may avoid the site entirely—but managers must be careful not to allow their inaction to invite illegality. If a site is very popular, an inadequate parking lot may induce dangerous and illegal parking rather than reduce crowd sizes.

More difficult problems arise when use must be rationed or metered. Rationing requires an agency not only to administer a system of allocation but to implement and enforce it effectively. For example, in recent years, a handful of high-profile national parks have implemented a reservation system for vehicles entering the parks. During busy times, visitors are required to access a website and reserve a time slot to gain entrance into the park. At other sites, key roads are closed to visitors entirely. At Zion National Park, only NPS buses are allowed on the canyon floor; visitors must leave their cars at large lots far away from the sites they wish to visit. Sometimes, even popular hiking trails or other attractions require an advance reservation. The harrowing Angels Landing trail in Zion, for example, became so crowded that park officials created a lottery system for access to the trail at peak times (see Figure 12.4).

Important Dates

DATES	INFORMATION
January 1, 2025 - January 20, 2025	Seasonal lottery for spring dates (hikes on March 1 – May 31, 2025)
January 25, 2025	Notification of results of spring seasonal lottery
February 28, 2025 - May 30, 2025	Day-before lotteries (apply one day in advance of desired hiking dates)
April 1, 2025 - April 20, 2025	Seasonal Lottery for summer dates (June 1 - August 31, 2025)
April 21, 2025 - April 24, 2025	Scheduled Trail Maintenance Closure - No permits issued for these dates
April 25, 2025	Notification of results of summer seasonal lottery
July 1, 2025 - July 20, 2025	Seasonal lottery for fall dates (hikes on September 1 - November 30, 2025)
July 25, 2025	Notification of results of fall seasonal lottery
October 1, 2025 - October 20, 2025	Seasonal lottery for winter dates (hikes on December 1, 2025 - February 28, 2026)
October 25, 2025	Notification of results of winter seasonal lottery

Figure 12.4 This screenshot from Recreation.gov shows the lottery timeline for the popular Angels Landing trail in Zion National Park. Source: Recreation.gov.

Systems like these pose thorny questions of policy. Should *all* slots be allocated online in advance, or should some slots be held for in-person visitors? If the latter, in what proportion? Should those who have previously climbed Angels Landing be deprioritized in the lottery system? Should foreign and domestic visitors be treated identically? Are online reservation systems fair in light of unequal access to the internet?

These and other questions can't be answered by simply consulting the law. The managing agency must exercise its discretion and make a carefully considered judgment. And while a site's staff can sometimes implement policies unilaterally without much opposition—adding a permit requirement for overnight camping, for example, might not raise anyone's hackles—in other instances their decisions will provoke loud disagreement. In these instances, land managers must act with caution and diplomacy.

The Colorado River within the Grand Canyon offers one such example. Rafting the River is a favorite activity within the Canyon, and most visitors who undertake it rely on professional outfitters to coordinate the entire one-to-two-week journey. As the number of these operators has grown, the remote Inner Canyon has become a congested corridor. The experience, once defined by unparalleled solitude, now involves queued fleets of rafts descending the various rapids, with groups frequently vying for prime sandbars, caves, and camping spots. The motorboats used by some outfitters for hurried guests—a necessary concession for those with limited time—introduce a jarring and unwelcome noise that shatters the silence for others.

As a consequence, the National Park Service has, for half a century now, limited the number of rafting excursions allowed within Grand Canyon National Park. The Service engages in a regular planning process that addresses the use of the Colorado River; in fact, the River is such a distinct resource within the Park that the Service maintains a Colorado River Management Plan apart from the General Management Plan that governs all other park operations. This Plan contains very specific limits on the number of individuals that can raft the river, the number of boats that can carry them, the number of those boats that may be motorized, and so forth. These numerical limits are not arbitrary but are the result of intensive study by the Park Service, as well as tense negotiation with rafting outfitters. Over the years, debates over rafting quotas have been so intense that they've led to litigation[6] and even attempts to pass specific congressional legislation to override the NPS's decisions.

To summarize, overcrowding creates all sorts of issues for land managers, from the prosaic matter of upkeep to practical challenges of implementation to thorny questions of policy and public relations. Yet again, the law supports agencies when they make careful, deliberate decisions; courts are generally willing to uphold their policies against lawsuits brought by disgruntled stakeholders.

Access

Just as overcrowding presents challenging questions of public lands management and philosophy, so also does the antecedent issue of access. Recreational use of the public lands depends on access—you can't ski, climb, or bike at a place you can't reach. At the most basic level, accessibility relates to transportation and its infrastructure; we label as accessible those places we can get to with relative ease using modern modes of transport. But today, accessibility as a concept goes much further. It addresses not only physical access but also social context—such as the invidious pressures that impair disadvantaged populations' access to certain universities or employers. Those same pressures can hinder people from visiting and using public lands.

As a matter of public ethics, the accessibility of public lands poses a complicated cluster of issues. This can be seen in ongoing debates about roads. Today we take for granted that we can drive nearly anywhere we wish to go; physical access to the public lands is provided most obviously by road vehicles, like cars and buses. Yet roadbuilding within the public lands has long been deeply contested. This is partly due to the environmental impacts associated with roads—the scars they leave on the landscape, the risks they pose to animals, the gateway they provide to additional development. But at a deeper level, roads symbolize the intractable tradeoff between *ease of access* and *wildness.*

Ease of access is an article of faith in American culture. Thanks to laws like the Americans with Disabilities Act (ADA), our built environment is generally able to accommodate visitors with a wide range of physical abilities. Many would argue that the public lands should offer the same openness and accommodation. Certainly, those with sensory or mobility impairments may be less likely than others to mountain bike or rock climb, but that makes it all the more important—or so the argument goes—that we have a network of roads to facilitate sightseeing, which is by far the most common recreational activity undertaken on many public lands. Sightseeing requires little of people; all one must do is look out the window of a car or bus. Nothing could be easier, and for some visitors to the public lands, any more intensive activity would be out of reach.

While this viewpoint is cogent and compassionate, there are counterarguments. They emphasize that the most obvious *means* by which most access is provided—roads—may facilitate far more harm than benefit. This is because roads bring all the destructive implements of modernity to the wildlands' doorstep. A road might seem like a narrow ribbon of land, cutting inconspicuously through a vast landscape, but roads have a way of expanding outward like a spider web, creating demand for more roads and more development—driveways, spurs, gas stations, restaurants, and eventually more and wider roads, until wildness is gone forever. And roads are permanent; for practical purposes, roads once built are not unbuilt.

BOX 12.4 Sax and Abbey on Accessibility

Joseph Sax (1936–2014) was a law professor, a prolific and poetic writer, and a prominent champion of the public lands. In his widely-acclaimed book *Mountains Without Handrails*, Sax defended the view that the national parks should be managed for a specific kind of activity which he termed "contemplative recreation." In the excerpt below, he takes up the question of accessibility. Although the passage reflects both the lexicon and categories of the time, it allows you to see how a widely-respected theorist of the recent past contended with the tension between access and development.

> The concern that has been expressed for the elderly and the infirm in debate over parkland developments must be taken with a measure of skepticism. People who were active when they were young ordinarily continue to be as active as they can when they get older, and those who are reluctant to leave their cars range widely across age groups. Neither the elderly nor the infirm, if they were active at other times, are in the forefront of those advocating intense development of parklands. Rather, those who urge development have put the elderly and the handicapped on their front line. I have myself climbed in Montana with a fifty-seven-year-old

> blind man, who was continuing—to the best of his ability—to pursue the kind of activity he enjoyed before his injury. And I have walked down and back up the Grand Canyon with a husband and wife in their late sixties, who at a slower pace were repeating adventures they had previously savored.
>
> Management committed to contemplative recreation should be just that, whether for the young and hardy or the old and infirm. One does not provide such an opportunity for older people or inexperienced visitors by building a highway to the top of a mountain. Rather we can assure that places that *are* accessible to them are not so deprived of their natural qualities as to put such an experience beyond their reach. If it were necessary to go into the rugged backcountry before finding a relatively undisturbed ecosystem, the lesson would be that we had too ruthlessly developed the more accessible places, not that still more places should be deprived of their complexity.[7]

Contrast Sax's nuanced argument with the coarser view expressed below by Edward Abbey (1927–1989), an American writer and naturalist perhaps best known for his novel *The Monkey Wrench Gang*. His book *Desert Solitaire* was an account of his years as a park ranger in Arches National Monument (now a National Park) in Utah. Full of riveting stories, the book includes Abbey's strident yet influential views on park roads:

> Industrial Tourism is a threat to the national parks. But the chief victims of the system are the motorized tourists. They are being robbed and robbing themselves. So long as they are unwilling to crawl out of their cars they will not discover the treasures of the national parks and will never escape the stress and turmoil of those urban-suburban complexes which they had hoped, presumably, to leave behind for a while.
>
> How to pry the tourists out of their automobiles, out of their back-breaking upholstered mechanized wheelchairs and onto their feet, onto the strange warmth and solidity of Mother Earth again? This is the problem which the Park Service should confront directly, not evasively, and which it cannot resolve by simply submitting and confirming to the automobile habit. The automobile, which began as a transportation convenience, has become a bloody tyrant (50,000 lives a year), and it is the responsibility of the Park Service, as well as that of everyone else concerned with preserving both wilderness and civilization, to begin a campaign of resistance. The automotive combine has almost succeeded in strangling our cities; we need not let it also destroy our national parks.[8]

Abbey went on to argue that there should be no more cars and no new roads in national parks, full stop.

Roads are merely the tip of the iceberg in debates about accessibility. In thinking about recreation in its many forms, land managers must wrestle not only with how to balance physical access with preservation but also with how to create a sense of belonging and ownership of the public lands among disadvantaged communities, who, historically, have been among the least likely to visit the public lands. Crafting such strategies might involve, for example, preserving and highlighting sites that have historical or cultural significance to a particular community; investing in marketing or educational campaigns that target groups otherwise unlikely to visit the public lands; or reducing the various charges and fees that can make the public lands too expensive for many would-be visitors.

Here is an example of how that can work. Shortly after his inauguration, President Biden issued Executive Order 13985 on "Advancing Racial Equity and Support for Underserved Communities Through the Federal Government." Among other things, the Order required federal agencies to examine "potential barriers that underserved communities and individuals may face" in taking advantage of government resources. In response to this Order, the U.S. Forest Service issued an Equity Action Plan in 2022. One of the priorities in that document was described as follows:

> **Promote Access to Recreation and Outdoor Experiences within Communities of Color and Socially Vulnerable Populations:** Promote research to understand who visitors are; how they recreate; who is not visiting; barriers to accessing recreation opportunities; economic, community, and personal benefits of recreation opportunities; and preferences regarding and satisfaction with recreation opportunities. Develop research-based strategies to increase visitation by communities of color. Promote use of the Native American Tourism and Improving Visitor Experience Act in support of Tribal culture and tourism enterprise.[9]

In the following year, the Forest Service began to implement that objective. It first took some general steps, like adding specialized employees, revising its communication materials, and analyzing ways that marginalized groups were experiencing barriers to public land use.[10] More specific actions followed. For example, the San Gabriel Mountains National Monument north of Los Angeles was one target of attention. This USFS property receives over 4 million recreational visits per year from Angelenos and others. Pursuant to the Equity Action Plan, the Service added bilingual staff rangers to support communication with the diverse local population. It also established a "Transit to Trails" shuttle to bring visitors into the Monument from transit centers in underserved communities in the Los Angeles area.[11]

Ultimately, debates over public lands accessibility are about competing values. The American commitment to ease of access—the belief that treasured places should be effortlessly reachable—is challenged by the reality that true wildness is defined by its resistance to domestication. As managers strive to ensure fair access for all communities and abilities, they must continuously confront this tradeoff: every measure taken to make the wilderness more accessible risks diminishing the very qualities of wildness that generations have sought to protect. The search for balance is not a simple calculation but an ongoing, delicate negotiation.

Concessioners

Finally, no discussion of recreation on public lands would be complete without some consideration of the role played by private enterprise. Many visitors to the public lands rely on outfitters, guides, tour companies, or countless other businesses to facilitate their recreational activities. But the functions of private business in public land recreation are far more pervasive than that. Anytime you visit a gift shop, a restaurant, a lodge, a campground, or a marina on the public lands, it's almost certain that you are dealing with a business rather than a federal agency. In some instances, these businesses operate with minimal oversight. But more commonly, businesses must obtain a land manager's permission before accessing the public lands for commercial purposes. The permission to operate a specific business is called a "concession," and the business receiving such permission is called a "concessioner" (or sometimes a "concessionaire"). This section will briefly describe how concessioners fit into the picture we've assembled of public lands recreation.

Concessioners provide a wide array of services across the public lands, but their role is most conspicuous in the National Park System, where they operate iconic lodges, gift shops, and camp stores; guide mountain climbers, river rafters, and backpackers; provide transportation by bus, bicycle, or ferry; and even offer educational programs for children and adults. The NPS website boasts that the agency "administers nearly 500 concession contracts that, in total, gross over $1 billion annually" and employ more than 25,000 people during peak seasons.[12] Concessioners' role is less pervasive but still significant on lands managed by the USFS, BLM, and FWS, comprising much of the day-to-day, public-facing service provided by these agencies. For example, concessioners operate facilities like campgrounds, picnic areas, and parking lots.

There is one category of public land on which you won't find concessioners; by now, you can probably guess which one: wilderness. The federal Wilderness Act prohibits "commercial activities" except where those activities are carried out under "valid existing rights," such as pre-existing grazing or mining rights.

Public land concessions raise several concerns. The first is about money. Since concessioners are, in essence, profiting from the use of a public asset, there will always be those who argue that such entities should return much of that profit to the public. Needless to say, this is an argument that concessioners themselves tend to resist. As a practical matter, there is no easy formula by which the proper value of a concessions contract can be priced, and in any event the federal government has historically had a difficult time maximizing income from its own land.[13]

BOX 12.5 Ski Resorts on the Public Lands

Although it's not obvious looking down from a chair lift, most ski resorts in the western U.S. make use of public lands. In 2017, Forest Service chief Tom Tidwell said the following in a speech he gave to the National Ski Area Association:

> The main way that Americans now use their national forests and grasslands is for outdoor recreation, and skiing is by far the most popular winter use. There are 122 ski areas on the national forests, and they provide about 60 percent of the total capacity for downhill skiing in the United States. Downhill skiing and snowboarding is the second most popular primary use of the national forests—after hiking. About 16 percent of our visitors come for downhill skiing and snowboarding. We get about 23 million skier visits per year, roughly half of all skier visits nationwide.
>
> All those visitors generate tremendous economic benefits. Ski areas on the National Forest System support about 64,000 full- and part-time jobs. Revenue to the federal Treasury from ski areas has been going up; in 2016, it reached the highest levels ever, more than $45 million. That's more revenue than from any other activity on the National Forest System—more than timber and all other recreation special uses combined. More importantly, these same ski areas

contribute to the national economy. They alone contribute about $2.7 billion annually to the gross domestic product, or about a quarter of the total contribution of the National Forest System.[14]

Ski resorts are lucrative, and one recurrent question has been whether the federal government is getting its due from resort operators. In fact, there is a special body of federal law devoted to the calculation of the rental charge that ski operators must pay to use Forest Service land, which is typically just a few percent of the resort's revenue.[15] Critics allege that wealthy operators like Vail Resorts are not paying the fair market rental value of the land, but the operators argue that the market value is not the appropriate point of reference because that number has been increased by the investments made by the resorts themselves—in other words, popular ski resorts drive up the market value of adjacent public lands.

The second concern is that certain concessioners are able to monopolize business opportunities on the public lands. Ordinarily, there is aggressive competition for good business opportunities, but historically the land management agencies have simply renewed their contracts with incumbent concessioners rather than solicit fresh bids. Over the years, Congress has tried to create a more competitive process for federal concessions,[16] and as a result some of the most prized contracts have changed hands in recent years. For example, in 2015, Aramark won the right to operate the marquee lodges inside Yosemite National Park, a contract that had previously been held by Delaware North. Ironically, however, Congress's efforts may have had the unintended consequence of benefitting the largest concessioners—companies like Aramark and Xanterra—which have ample experience navigating the complex federal contracting process.

At the end of the day, the relationship between the concessioners, land managers, and the public is a tangled one. Land managers rely heavily on concessioners to provide certain services to visitors. Even as critics claim that concessioners benefit inappropriately from public resources, visitors typically welcome the services they provide, and are sometimes unable to distinguish between private and public services during their visits to public lands.

Conclusion

Recreation is the basis of most Americans' use and awareness of public lands. Yet unlike other major uses, recreation is not directly addressed by a distinct body of federal land law. This leaves the land management agencies with the freedom to craft recreational policies to their own liking, as long as those policies are thoughtfully drafted and informed by public comments.

But the lack of law should not be confused with a lack of public attention. People care deeply about their favored modes of recreation, as do the businesses that sustain them. Constructing effective policies that accommodate visitors while protecting the land (and enabling other activities) requires care and caution. Whether building a road, managing a busy trail, or contracting with an outfitter, land managers confront numerous public and practical pressures. And no sooner is a policy complete than it must be reconsidered, for Americans' recreational preferences change as quickly as the shadows of sunset at Horseshoe Bend.

Study Questions

1. How do the Wilderness Act's prohibitions on "mechanical transport" and "installations" create policy challenges for public land managers?
2. In the absence of more specific laws, how do the multiple use (BLM/Forest Service) and dominant use (NPS/FWS) shape decisions about recreational policy?
3. What ethical or equity concerns arise from managing overcrowding through online, reservation-based rationing systems for access to popular public land sites?

Resources for Further Study

Abbey, Edward. *Desert Solitaire* (New York, NY: McGraw-Hill, 1968).

Bradshaw, Karen, and Caitlin Doak. "Making Recreation on Public Lands More Accessible." *Notre Dame Law Review Reflections* 97 (2022): 33–51.

Kenyon, Mark. *That Wild Country: An Epic Journey Through the Past, Present, and Future of America's Public Lands* (New York, NY: Little A, 2019).

Miles, John C. *Wilderness in National Parks: Playground or Preserve* (Seattle, WA: University of Washington Press, 2009).

Sellars, Richard. *Preserving Nature in the National Parks: A History* (New Haven, CT: Yale University Press, 1997).

Notes

1 William C. Tweed, *Uncertain Path: A Search for the Future of National Parks* (Berkeley, CA: University of California Press, 2010), 172–3.

2 16 U.S.C. § 1131(c).

3 To remind yourself about these different standards, flip back to Chapter Five, which first mentions the distinction between multiple use and dominant use agencies.

4 U.S. Dept. of the Interior, Bureau of Land Management, *Rock Springs Field Office: Draft Resource Management Plan Revision and Draft Environmental Impact Statement* (May 2023), pp. 3–23.

5 By contrast, the climbing community recognizes that power drills are "motorized equipment" within the meaning of the Wilderness Act. Consequently, climbers in wilderness areas generally use hand drills rather than power drills to secure their climbing bolts into rock faces.

6 See, e.g., River Runners for Wilderness v. Martin, 593 F.3d 1064 (9th Cir., 2010).

7 Joseph L. Sax, *Mountains Without Handrails: Reflections on the National Parks* (Ann Arbor, MI: University of Michigan Press, 1980), p. 80. Emphasis in original.

8 Edward Abbey, *Desert Solitaire* (New York, NY: McGraw-Hill, 1968), 49.

9 U.S. Forest Service, *Equity Action Plan* (July 2022), 4.

10 U.S. Forest Service, *Equity Action Plan 2023–24* (July 2023), 13.

11 U.S. Forest Service, *Investments in San Gabriel Mountains National Monument* (2024).

12 "Ensuring High Quality Visitor Services," https://www.nps.gov/subjects/concessions/index.htm.

13 Bruce Huber, "The Fair Market Value of Public Resources," *California Law Review* 103 (2015): 1515–1559.

14 Tom Tidwell, "Ski Areas: An Enduring Partnership," speech delivered May 5, 2017, available at https://www.fs.usda.gov/speeches/ski-areas-enduring-partnership.

15 See, e.g., 16 U.S.C. § 497(c).

16 See, e.g., The National Parks Omnibus Management Act of 1998, Public Law No. 105–391.

13 Mining and Energy Development on the Public Lands

Introduction

When you enter a national park, it can feel like you're entering a museum of nature. There are signposted trails, grand vistas, and gawking tourists. Heavy industry is nowhere to be seen. Yet on other public lands, the opposite is true. You may come across a large-scale mining operation, with massive earth-moving trucks and cranes, or an oil and gas field with wellheads and pumpjacks—and not a tourist in sight.

To prospectors and extractive industries, the federal public lands are more likely to be regarded as vast storehouses of mineral wealth than as scenic or recreational sites. Resource extraction is an enduring reality on these lands, and overseeing the extractive process (and its aftermath) is an important function of the public land manager. The legal pathways by which minerals[1] may be discovered, extracted, and removed from the land are the primary subject of this chapter. Yet as attitudes toward fossil fuels have changed, so also have priorities concerning the use of public lands. Thus this chapter will also discuss how declining demand for coal and increasing demand for renewable energy—and the land and minerals it requires—are shaping the management of the public lands.

The chapter will begin with some general background and then will turn to discuss *mineral location* and *mineral leasing*, the two principal ways that resources may be extracted from public lands. After that, we'll turn to renewables. The chapter will end with a brief discussion about the process of reclaiming land that has been damaged by resource extraction.

Background

Let's begin with a bit of a refresher. You've already learned about the General Mining Act of 1872 (also known simply as the "Mining Act") and the myriad mining claims that dot the western public lands (Chapter Two). Chapter Three discussed the concept of mineral leasing, which applies to fossil fuels, and indicated that some lands have been withdrawn from the operation of the Mining Act and the mineral leasing laws. The BLM, the leading agency in this chapter, was covered in Chapter Eight, which also provided more detail about mining, mineral leasing, and the planning processes that undergird them both. Federal law gives BLM the job of managing subsurface minerals on *all* federal lands, regardless of which agency manages the surface. Roughly 1,000 of the agency's more than 10,000 full-time employees are involved in energy or minerals management.

DOI: 10.4324/9781003392514-13

With that background in mind, let's dig deeper into the distinction between the Mining Act's system of *mineral location* and the Mineral Leasing Act's system of *mineral leasing*. These are two entirely distinct ways of allocating mineral resources, and they are quite different.

Mineral location is a system in which prospectors can enter public land and claim valuable mineral deposits simply by *locating* them. It's a "bottom-up" system that gives prospectors a wide-ranging right of self-initiation across vast amounts of public land. The minerals to which this system applies are called "locatable" minerals. The Mining Act of 1872 declared that, for lands open to mining, anyone can enter to prospect at any time. If they find a valuable mineral, they are entitled to claim it for themselves. That's why it's "bottom-up": regular people can enter the land and decide where to dig, and even the mighty federal government is obliged to respect their claims.

Mineral leasing, by contrast, is more of a "top-down," government-run system that applies to oil, natural gas, and coal—that is, the fossil fuels—as well as several other commercially important resources such as phosphates, sodium, sulfur, and potassium. For leasing to occur, land must be nominated by an interested party or by the managing agency. If the agency supports the nomination, it places the mineral rights up for lease in a competitive process that often attracts multiple parties. The winning bidder acquires the right to develop the leasable mineral, subject to a royalty payment to the U.S. government. The parameters of the process are fully spelled out by law and regulation, in true top-down fashion.

The next several sections will provide more detail about the location and leasing systems, but there is one more mineral disposal system you should know about. It's called the *sale system*, and it applies primarily to common minerals, such as sand, stone, and gravel. In certain instances, federal land managers may sell (or even give away) such minerals in a contractual process that is considerably more straightforward than either leasing or location.

BOX 13.1 What Exactly is a Locatable Mineral?

Which resources are *locatable*, and which are *leasable*? The leasable resources tend to be relatively widespread energy or fertilizer minerals, while locatable resources are rare and valuable substances, such as precious metals. But as a matter of law, *locatable* minerals are simply those that aren't *leasable* or *saleable*. It's a category that is defined negatively.

Here's how that works. By its terms, the Mining Act applies to "veins or lodes of quartz or other rock in place bearing gold, silver, cinnabar [mercury], lead, tin, copper or other valuable deposits." Most of those words are fairly precise, but what about "other valuable deposits"? That phrase is quite broad. Over the years, Congress removed certain substances from the operation of the Mining Act by making them either leasable (e.g., coal, oil, and natural gas) or saleable (e.g., sand, stone, and gravel). Every other substance remains locatable as long as it falls within the scope of the Mining Act.

There are plenty of substances out there that might constitute a "valuable deposit," and they've given rise to no small amount of litigation. What about groundwater? Stalagmites and stalactites? Quartzite (so-called "Fool's Gold")? Fossils? Moreover, what if there is only a trace amount of a certain mineral? With modern gold mining techniques, for example, a vein containing just 0.01 *ounces* of gold per *ton* of ore might be profitable if the market price of gold is high. Should such a vein be considered locatable, and should this determination change with the market price of gold? These questions and others like them have been raised time and time again. Courts do their best to interpret the Mining Act and apply it to situations that would have been unimaginable to its authors in 1872.

The General Mining Act and the Location System

At this very moment, you may enter the public lands and stake out a mining claim. To assert the claim, you need only mark it. The valuable minerals you find there are yours to keep, for free—"a benefit not available anywhere else on earth," as one commentator notes.[2] Even if you don't find valuable minerals, the federal government can only eject you after going through a lengthy administrative process to void your claim—at which point you may, if you wish, simply stake another claim and start the process again without penalty. Until 1994, you could have gone one large step further and, by completing some paperwork, acquired outright title (a "land patent") to your claim. The land would become yours to keep. Does this sound too good to be true? We have left out a few important caveats—for example, the land must be open for mining—but the basic scheme is remarkably (and perhaps gallingly) generous.

This giveaway comes to us thanks to the Mining Act, which though over 150 years old has never been repealed and is very much still in effect. At its core is the so-called "right to mine": the idea that on unreserved public lands anyone may search for minerals and upon locating a deposit may stake a claim to that deposit within the parameters of the Act. Newcomers to the topic often find the right to mine to be one of the strangest and most interesting aspects of public land history and law. Practically speaking, the right to mine means that the BLM and USFS lack the legal authority to stop prospectors from exploring for, discovering, and developing minerals on much of the land they manage. In a moment, we'll examine how these agencies can regulate the mining process and even withdraw land from mining altogether. The key point, however, is that even an unpatented mining claim constitutes a powerful property right, one that is strikingly easy to acquire.

BOX 13.2 Mining Act Abuse

Abuse of the Mining Act was commonplace through much of the 1900s as thousands of claimants took advantage of the Act's liberal terms. A federal investigation in 1990 found that people used mining claims to grow marijuana, build vacation cabins, and even store toxic chemicals. "The most common abuse involves the construction of houses and cabins, many with 'amenities such as gazebos, garages, greenhouses and satellite television dishes,'" reported *The Washington Post*, quoting the federal report.[3] "All these claim holders live rent-free on public land." Other people try to turn a quick profit by staking claims where federal projects are planned. The federal investigation noted that one claimant "recently claimed 540 acres of Nevada desert that happened to be part of the site of a proposed nuclear waste repository to be operated by the Energy Department at Yucca Mountain. Although the repository plans were well known at the time the claims were filed, last year the department was forced to buy out the claim holder for $249,500."[4]

Why is the Mining Act so conducive to fraudulent mining claims? You might think that government agents could simply walk up to illegal claimants and order them to leave, but it's not that simple. The public lands are vast and agency personnel are few—and agency staffers have far more to do than evaluate the adequacy of mining claims. Even when the government is alerted to a suspicious claim, the process of ejecting an illegal claimant is complicated. The burden of proof is on the agency: the law requires the agency to prove that a claim is invalid, rather than the claimant to prove that the claim is valid. In effect, claimholders can simply stay put and require the government to go through a time-consuming and expensive process to demonstrate the absence of valuable minerals. And there is nothing punishable about filing an illegitimate claim. If a claimant is ejected from a bad mining claim, there's nothing to prevent them from simply staking out another claim on the very same day.

Consistent with nineteenth-century mining practices, the Mining Act creates two kinds of claims—*lode* claims and *placer* claims—with different rules for each. In mining jargon, a lode is a well-defined vein of mineral ore, still held in place in its native rock. A placer deposit, by contrast, is one in which the mineral of interest has been worn down and scattered, usually by water or ice, such that the mineral is now found amidst a bed of sand, gravel, or loose stone. Lode claims are limited by law to 1,500 feet in length and 600 feet in width, an area which totals a bit more than 10 acres. Placer claims have a maximum size of 20 acres and are usually described with reference to the formal boundaries of a federal land survey (see Figure 13.1).

As you might imagine, modern mining operations are much larger than 10 or 20 acres. To develop new mines, today's mining firms must assemble multiple claims, and often must also acquire rights to intermingled state or private land. They also have to figure out how to dispose of waste rock, known as "tailings." Modern mines excavate enormous amounts of earth, processing tons of ore to extract ounces of the desired

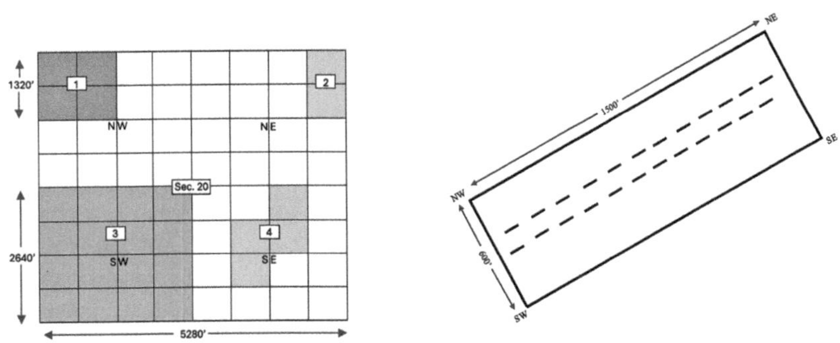

Figure 13.1 Examples of placer (left) and lode (right) mining claims. The diagram on the left represents a hypothetical Section 20, divided into 64 10-acre cells. Placer claims may ordinarily be up to 20 acres in size (two cells on this diagram), as is claim #2 on the top right. Claims #1 and #4 are each 40 acres, while claim #3 is 160 acres. These claims would require several claimants to join together to create an "associational" claim. The diagram on the right shows a sample lode claim, a rectangle of 600 by 1500 feet oriented along the lode itself, represented by the dotted lines. Source: BLM.

mineral. Tailings are piled high on adjacent land. An open-pit mine, thousands of feet wide, might require a tailings pile of several square *miles.* [5] The Mining Act's only allowance for off-site disposal, however, grants miners a mere five acres for use as a "mill site."

To accommodate large tailings piles, federal agencies have had to bend the Mining Act to its breaking point, allowing mining firms to exceed the Act's outdated strictures. But some Interior Department lawyers, prompted by persistent litigants, have started to require land management agencies to play by the Act's rules, perhaps gambling that Congress will finally update federal mining law if only it can be made aware of how archaic the law actually is.[6]

How much mining actually takes place under the location system? And exactly how valuable are the minerals that we allow mining companies to acquire for free under the Mining Act? There is no easy way to find out. The Mining Act does not require claimants to report the type or quantity of minerals they obtain from federal lands, a fact that has drawn much criticism. The BLM tracks the number of active mining claims on public lands—nearly 500,000 as of 2022—but "active claims" are simply those for which the owner has paid an annual maintenance fee, required to hold the claim. Actual commercial mining operations are far less numerous; the BLM reports that as of April 2022 there were 578 active mining plans of operation and another 867 active mining notices on federal land.[7] Again, the amount of production from those mines is known only to their proprietors. Over the years, many members of Congress have tried to amend the law to require a simple disclosure—even a rough estimate—of the value of extracted minerals. Thus far, the mining industry has rebuffed these attempts. It is clear, however, that mining operations are worth billions of dollars per year and provide not only an important source of local employment but also vital contributions to the national economy.

For the moment, the location system lives on despite obvious signs of obsolescence. Its underlying principle of self-initiation imposes significant constraints on land management agencies. As we'll see, the BLM and USFS retain some regulatory leverage over mining companies, but in an important respect, the "right to mine" gives those companies the first move. The federal agencies are legally bound to accommodate and work around valid claims, even as they strive to contain adverse impacts.

The Mineral Leasing Act and the Leasing System

Congress passed the Mineral Leasing Act in 1920 to address the fact that valuable fossil fuel reserves were simply being given away under the disposal laws. The 1920 legislation embraced a new system of leasing for resources like coal, oil, and gas. The leasing system requires lessors to pay a substantial royalty to the federal government. Because leases are contracts, the federal government can insert provisions into lease agreements during their negotiation, enabling the land management agency to require more from lessors than the bare statute would demand. For example, agencies now include reclamation requirements in these agreements, so that mining companies are bound by contract—if not by the Leasing Act itself—to restore the land surface upon the expiration of a lease. Crucially, leases may only be obtained on lands the government affirmatively determines to be appropriate for leasing. There is no right of self-initiation for leasable minerals—no "right to lease" equivalent to the "right to mine." This system has evolved over the years, but these core attributes remain intact today. Let's look more closely at how leasing works for oil, gas, and coal.[8]

Oil and Gas

The concept of the mineral lease first emerged in the oil and gas industry. Since its earliest days, the industry understood that it would be needlessly expensive to purchase land overlying an oil field; only a limited right to access oil and gas was required. But negotiating for that right was a challenge. Neither the surface owner nor the oil company knew *whether* or *how much* oil or gas was present. Oil firms were reluctant to overpay, knowing that most drilled wells are not commercially viable; only a few strike "pay dirt." Landowners were afraid of selling oil and gas rights too inexpensively; what if they were sitting on a huge oil field?

The lease was the solution.[9] In an oil and gas lease, the producer pays the landowner a royalty—conventionally one-eighth—on all oil or gas actually produced. The fixed royalty ensures that both the landowner and the oil company benefit if oil is found. Lease agreements also include a one-time "bonus" payment from the producer to the landowner; this payment secures the producer's exclusive right to begin production on that property within a set period (usually 1 to 10 years). While royalty rates are standardized—that is, firms don't usually compete on royalties—bonus payments fluctuate enormously, allowing oil firms to compete for valuable tracts. A lucky landowner might receive a bonus payment worth millions, followed later by lucrative royalties.

The Mineral Leasing Act mirrors this structure. Land is put up for lease in a competitive bidding process, typically conducted every quarter at the BLM state offices. Lease tracts ordinarily cannot exceed 2,560 acres (four square miles). The winning producer's bid serves the same purpose as the bonus payment, granting that party the exclusive right to explore the tract for oil and gas during the 10-year primary term. If oil or gas is discovered before the term ends, the developer is entitled to continue production as long as any wells on that tract remain profitable. Although there is sometimes fierce competition, many tracts go unsold and are offered for sale again in the following quarter. The Act establishes a royalty (usually one-eighth or 12.5%) payable for all oil and gas produced; although the royalty is payable to the federal government, federal law actually diverts half of this revenue to the state in which production occurs. Politically, this incentivizes state representatives to lobby for production on federal lands within state boundaries.

Sometimes, even leased tracts go undeveloped for many years. This can occur when oil and gas firms assemble large portfolios of land to stymie competitors, or perhaps to conduct widespread exploratory activities before deciding when and where to actually drill. Producers don't often publicize their intentions along the way. This creates a curious political dynamic when public lands are involved. In presidential campaigns, for example, one candidate might accuse an incumbent president of stifling oil and gas production from public lands, while a competing candidate might observe that oil and gas producers ought not be able to lease more land until they start developing the leases they've already obtained.

Coal

Coal is abundant on public land, especially in northeastern Wyoming, southern Montana, and parts of Colorado and Utah. The Powder River Basin in Wyoming and Montana, much of which is managed by the BLM and USFS, is one of the largest coal reserves on the planet (see Figure 13.2). Coal from the Basin is shipped by train not

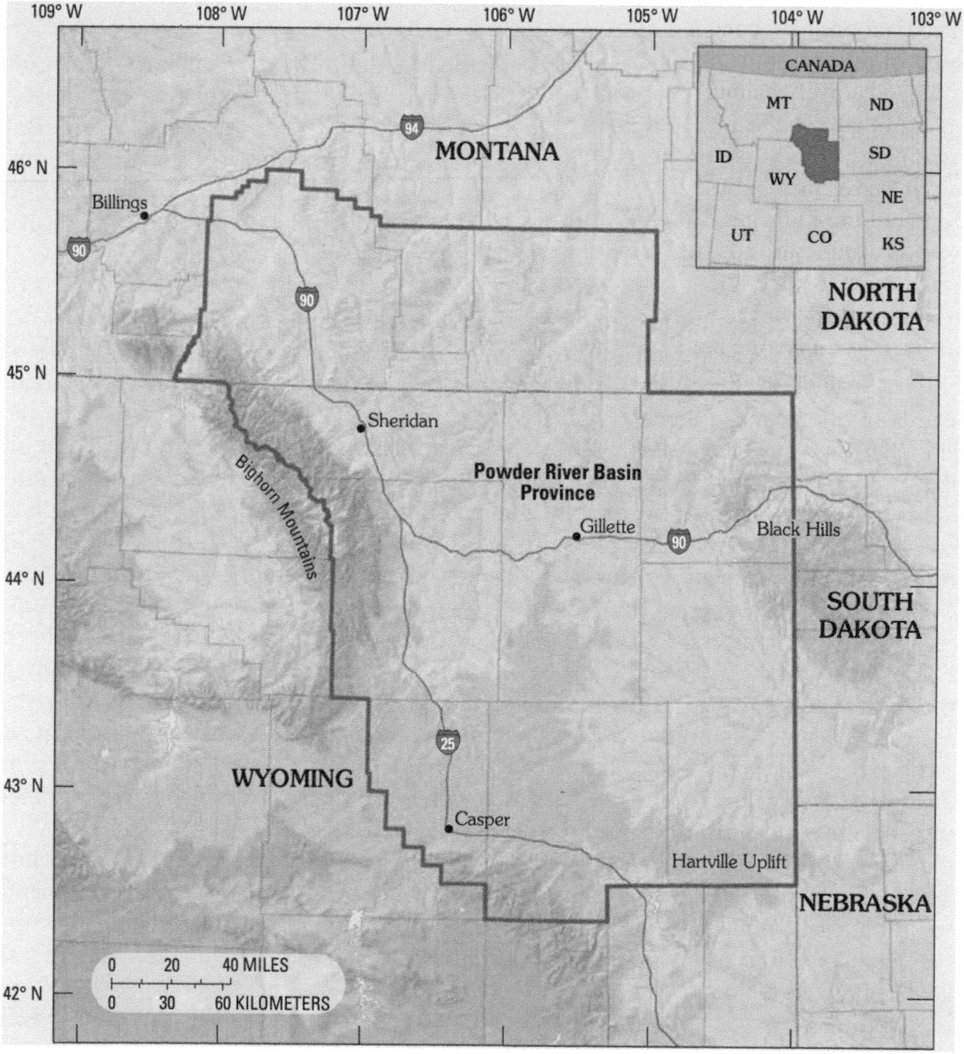

Figure 13.2 The Powder River Basin. Source: U.S. Geological Survey.[10]

only to coal-fired power plants across the U.S. but also to ports on the Pacific Coast, which send it as far as South Korea. Coal production in the Basin has been declining since its peak in 2008, but Wyoming still produces far more coal than any other U.S. state. Nearly all of it comes from the Powder River Basin.

The coal leasing process is procedurally straightforward yet deeply fraught. The process begins when coal firms nominate tracts of land they wish to mine.[11] The BLM then holds a competitive lease sale, with a price floor set at the fair market value of the coal. The winning bidder acquires the right to mine coal from the leased tract, with a minimum royalty of one-eighth the value of the coal. So far, so good.

The problem is in that little phrase "fair market value." In theory, a competitive sale should ensure that the government receives top dollar for its coal, and the price floor should act as a safety mechanism to guarantee that outcome. Yet numerous investigations have concluded that, over the years, the BLM's coal sale process has left vast sums of money on the table.[12] This is because most lease sales are not actually competitive at all. The overwhelming majority attract only one bidder: the party that nominated the tract in the first place. Why? It's because the nominated tract is usually adjacent to a pre-existing mine owned by that bidder. In this fashion, existing mining firms can extend their reach across the public lands, growing their operations incrementally without fear of competition, knowing that the cost to expand a pre-existing mine is far less than a competitor would have to pay to initiate a new operation.[13] But why doesn't the government simply set the price floor at a higher point? The law gives the agency enormous discretion to determine what the fair market value is, and in most instances, the agency seems more inclined to satisfy the industry's demand for coal than to maximize its income.

BLM's coal leasing practices have been under fire for other reasons as well. Some allege that BLM should do more to account for the costs of climate change in its assessment of fair market value; others assert that U.S. coal should not be sold into foreign markets. BLM's responses to such charges vary with each new administration. But one thing is clear: U.S. coal production, both on and off the public lands, is in the midst of a long-term decline (see Figure 13.3). This is partly a function of increasing concern about climate change and air pollution, but the more immediate reason for coal's slide is the relative abundance of cheap domestic natural gas over the past several decades. That has led the electric power industry to replace aging coal-fired power plants with new gas-fired ones.

Regulating Mineral Extraction on Public Lands

There are various ways the government regulates mines on public land. First, land management agencies can limit which lands are open to mining in the first place. Second, those agencies, along with other state and federal agencies, can regulate mining practices and techniques. Finally, regulators have made rules about mine closure; these rules require mine operators to reclaim mined land, if possible, or to mitigate the harms associated with defunct mines.

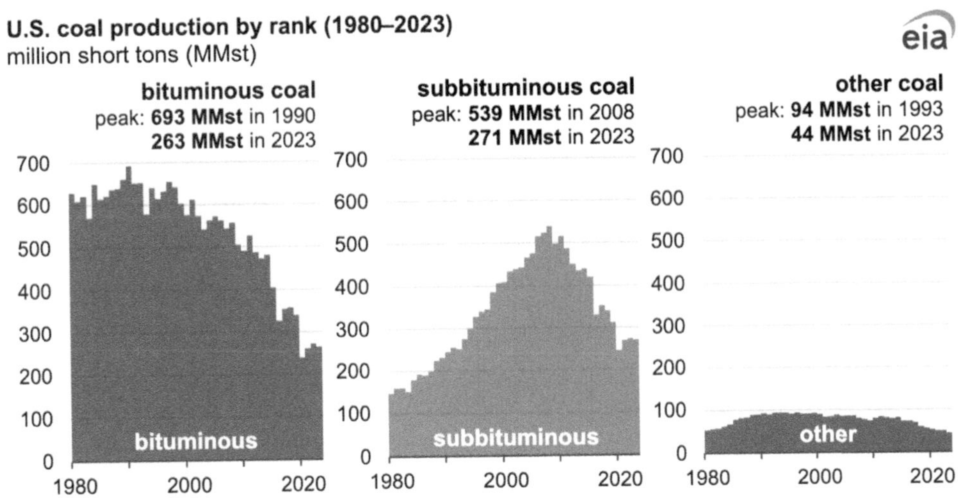

Figure 13.3 U.S. Coal Production, 1980–2023. Source: U.S. Energy Information Administration.

What Land is Open to Mineral Extraction?

As you learned in Chapter Three, there were nasty political fights in the early 1900s about the government's authority to withdraw lands from settlement. Now, thanks to FLPMA, that authority is well established. The power to make decisions about public land use is now a significant prize for the political party that holds the U.S. presidency. With little more than the stroke of a pen, presidents (or the officers they appoint) can decide which lands are or are not available under the mineral disposal laws. See Figure 13.4 for an example of a recent withdrawal.

Under the Mining Act, of course, the default is that lands are open to mineral location unless they have been withdrawn. Under the Leasing Act, it's the opposite;

DEPARTMENT OF THE INTERIOR

Bureau of Land Management

[223.LLAK941200.L1440000.ET0000; AA-95542]

Public Land Order No. 7922; Mendenhall Glacier Recreation Area; Alaska

AGENCY: Bureau of Land Management, Interior.

ACTION: Public Land Order

SUMMARY: This Public Land Order (PLO) withdraws approximately 4,560 acres of National Forest System Lands near Juneau, Alaska, from location and entry under the United States mining laws, and from leasing under the mineral leasing laws, for a 20-year period, subject to valid existing rights, to protect the recreational and natural resource values of the lands adjacent to the Mendenhall Glacier Recreation Area as the Mendenhall Glacier recedes.

DATES: This PLO takes effect on June 2, 2023.

Figure 13.4 An example of a land withdrawal. This excerpt is from a 2023 public land order withdrawing land from mining claims and mineral leasing. The withdrawal is relatively small, covering less than eight square miles, and protects an area newly exposed by a receding glacier. Source: 88 Federal Register 36333 (June 2, 2023).

land is not available for leasing until an agency makes it so. Most of the time, both of these decisions—whether to withdraw land from Mining Act location or to make it available for leasing—are made pursuant to the BLM's resource management planning process rather than direct presidential edict.

BOX 13.3 Minerals on Tribal Reservations

What about minerals on Native American reservations? It's complicated. Because reservation land is held in trust for tribes by the federal government through the Bureau of Indian Affairs (BIA), tribes must obtain the permission of the BIA before developing or selling minerals. Adding further confusion, the federal government sometimes owns outright the mineral estate beneath portions of a reservation (i.e., the tribe owns only the land's surface, not the subsurface). All of this creates a cumbersome bureaucratic process which can deprive tribes of the ability to make independent decisions about mineral management. There have been several high-profile instances in which the BIA was found liable, to the tune of billions of dollars, for failing to negotiate adequately on tribes' behalf. Although this book focuses on public lands and not tribal reservations, you should nonetheless be aware that the Interior Department plays a crucial and contested role in the disposition of many of the minerals underlying Native American reservations.

To get a sense for how this process works, let's consider the part of southwestern Wyoming managed by the Rock Springs Field Office of the BLM. As you can see in Figure 13.5, the BLM manages most of the acreage within this region.

Like all BLM field offices, Rock Springs is required to update its Resource Management Plan (RMP) every several decades. It completed an update in 2024 and incorporated a thorough NEPA analysis into its lengthy report. It considered several different alternatives with respect to mineral entry. Alternative A is the status quo. Alternative B is a very resource-protective alternative. Alternative C is the opposite; it favors resource exploitation. In Figure 13.6, you can see that Alternatives B and C differ considerably with respect to the acreage that would be withdrawn from mineral entry (shown in green). Alternative C would actually open to entry several areas that are presently withdrawn, while Alternative B would withdraw vast portions of the region. We could display a very similar set of maps showing areas available to oil and gas leasing under the same alternatives.

In short, the availability of public lands under the mineral disposal laws is always changing. Withdrawals are a regular occurrence. If you listen closely to news about the public lands, you'll frequently hear stories about thus-and-such area being withdrawn from, or opened to, certain kinds of mineral uses. Obviously certain areas are more contested than others; many portions of the public lands, like national parks and wilderness areas, have been permanently withdrawn. Bear in mind also that nearly every withdrawal is subject to valid existing rights, meaning that properly established leases or mining claims survive even if the land on which they lie is withdrawn. For this reason, active mineral extraction sometimes occurs on land that is formally withdrawn.

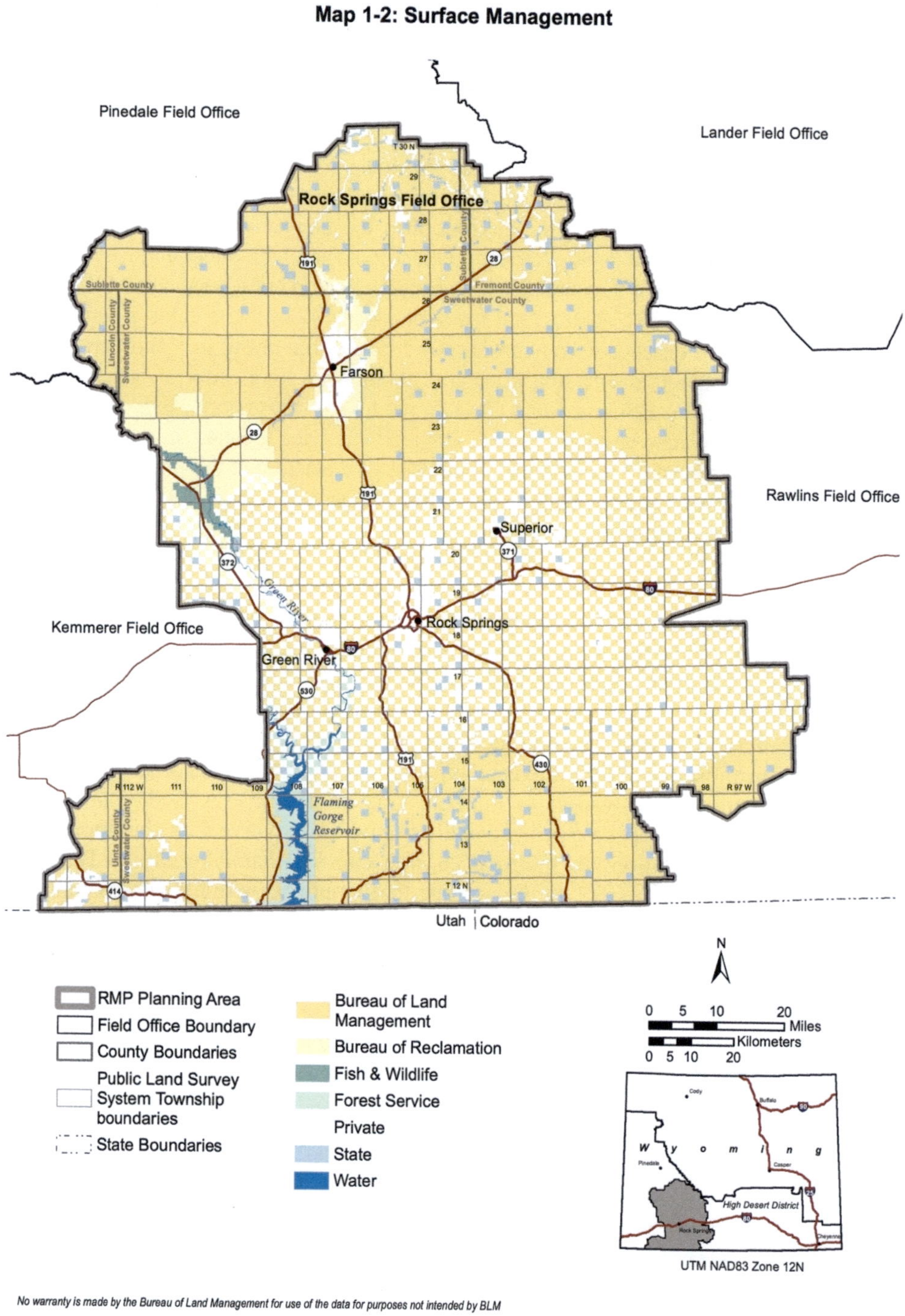

Figure 13.5 Surface management authority within the Rock Springs area. Source: BLM.[14]

Map 2-2: Proposed Withdrawal from Mineral Entry Alternative B

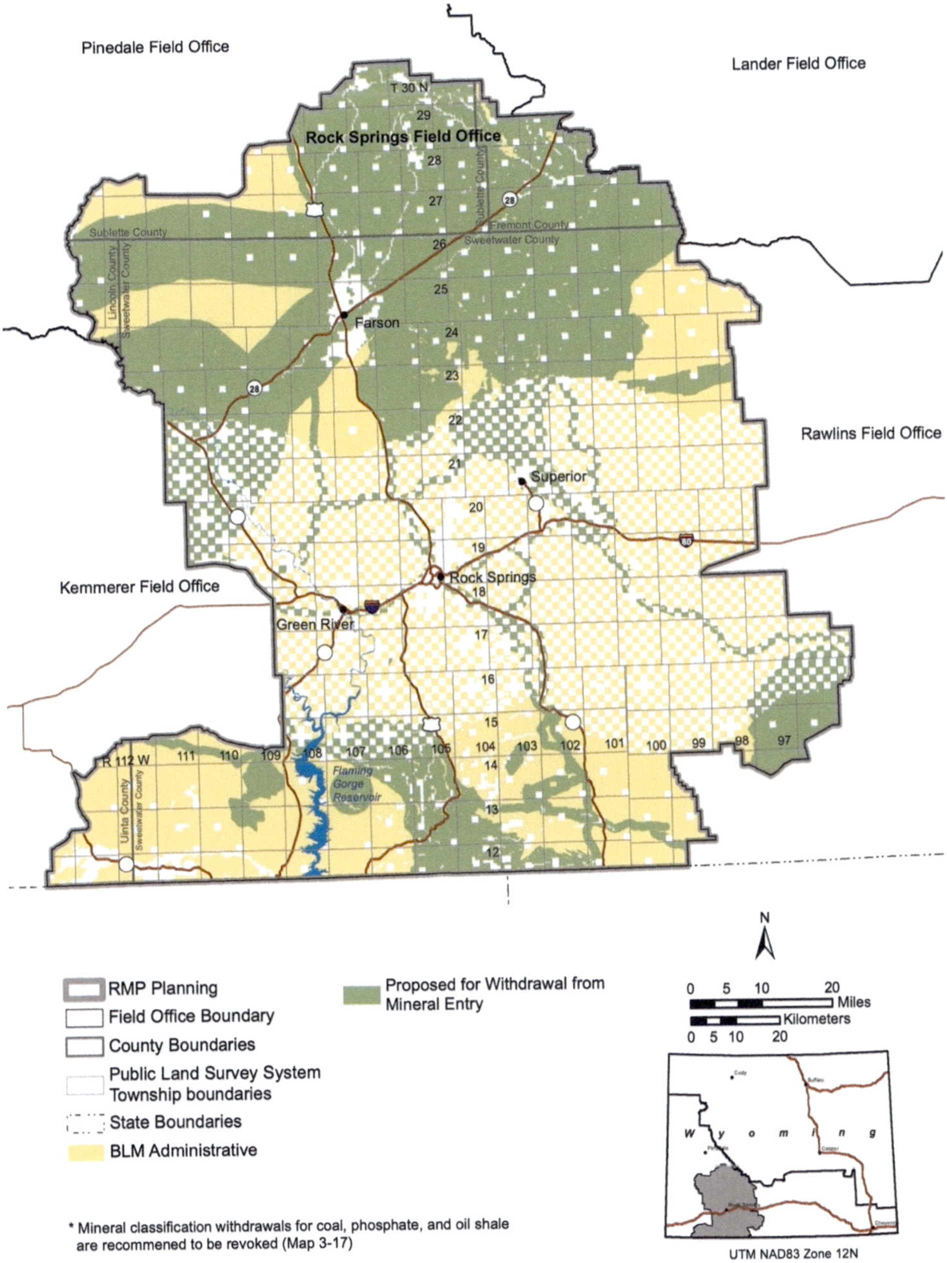

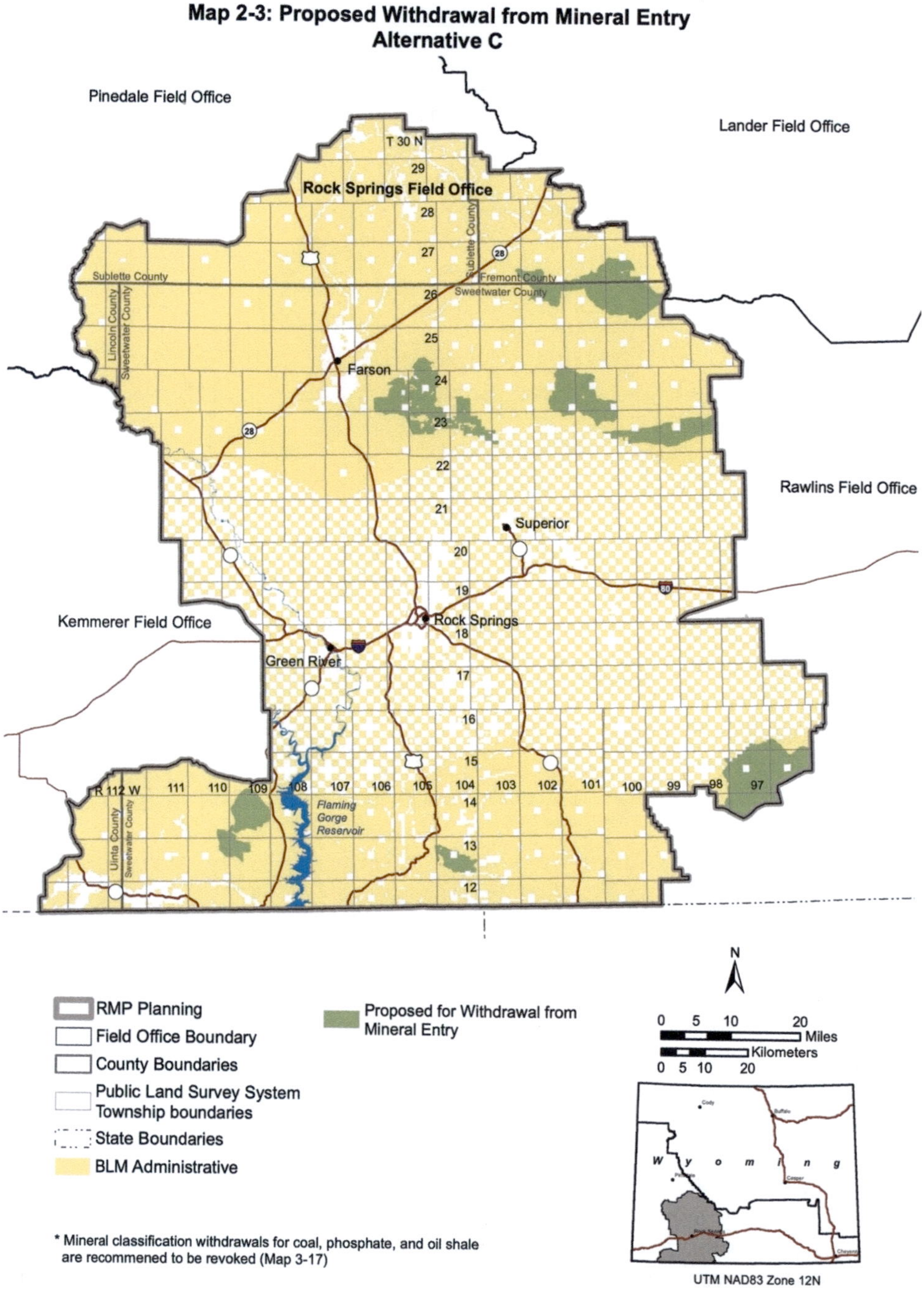

Figure 13.6 A comparison of two alternatives for mineral withdrawal in the Rock Springs RMP; areas proposed for withdrawal are shown in green. Source: BLM.[15]

Regulating Mining and Extractive Processes

Mineral development provides us with natural resources that undergird countless aspects of modern life. But it also degrades the environment. The more we learn about mining, the more harms we discover. One of the most prevalent harms associated with mining is called "acid mine drainage" (AMD). It occurs in various mining contexts, such as when rainwater is acidified by newly-exposed rock and flows into a nearby stream. It doesn't take much acidification to kill a stream habitat. Similarly, water that collects in mining pits can contaminate groundwater and migrate into aquifers several miles away, leaving the water unusable for agriculture or for human consumption. Moreover, some mining processes rely on toxic substances, such as when cyanide is used to leach gold out of ore. Needless to say, the containment of poisonous mining residues is a high priority for land use policy.

But balancing the necessity of extractive activities with their careful management has been an exceedingly difficult task for land managers. This balancing act is especially tricky under the Mining Act because its "right to mine" is sacred to the mining industry. When it passed FLPMA in 1976, Congress seemed to beef up the Interior Department's authority to regulate mines, but notice how it tried to thread the needle:

> Except as provided … in the last sentence of this paragraph, no provision of this section or any other section of this Act shall in any way amend the Mining Law of 1872 or impair the rights of any locators or claims under that Act, including, but not limited to, rights of ingress and egress. In managing the public lands the Secretary shall, by regulation or otherwise, take any action necessary to prevent unnecessary or undue degradation of the lands. 43 U.S.C. § 1732(b).

The unwieldy first sentence seems to suggest that the second sentence *may* actually amend the Mining Law and impair the rights of claimants, but the Secretary's new authority extends only to the prevention of unnecessary or undue degradation. That phrase—unnecessary or undue degradation—has become so iconic within public lands circles that it has its own acronym, UUD. The BLM has struggled to implement the "UUD mandate" and in particular has waffled on the question of whether the mandate allows the agency to deny outright a miner's ability to develop a particular claim, or whether it merely allows the agency to insist upon certain environmental precautions and practices. To this day, the BLM has never denied a mining proposal on the grounds that it was inconsistent with its obligations under the UUD provision above.

So where does that leave us? The BLM presently requires "plans of operation" to be submitted for mines that will disturb the land's surface. Its review of these plans triggers NEPA review, meaning that the agency may need to complete an EIS even if it can't or won't reject the mining plan. The plan may also trigger ESA consultation or permitting processes outside the BLM's authority, such as under the Clean Water Act. Additionally, for mining claims located on National Forest land, the USFS retains regulatory authority over the management of the land surface, and mine operators must submit plans to the USFS if their operations could cause a "significant disturbance of surface resources."[16] Those plans must address environmental protection, too. The two agencies typically work together when a mine implicates both agencies' resources, even though the legal mandates for the two agencies are somewhat different.

BLM's authority to impose environmental protections on oil and gas operations is more straightforward, in part because, as we've said, there is no "right to lease" equivalent to the "right to mine." When a lease is issued, the BLM may include provisions requiring certain protective measures, which then become binding upon the lessee just like an ordinary contract. A typical federal oil and gas lease will, for example, require the lessee to conduct operations "in a manner that minimizes adverse impacts to the land, air, and water, to cultural, biological, visual, and other resources, and to other land uses or users." The lease generally empowers the government to specify the measures required to accomplish that provision.

BOX 13.4 Flaring Methane on the Public Lands

Aside from its ability to insert protective provisions in oil and gas leases, the BLM also has the authority under the Mineral Leasing Act to issue regulations that apply to all oil and gas operators. For example, the Act requires that "[a]ll leases of lands containing oil or gas ... shall be subject to the condition that the lessee will ... use all reasonable precautions to prevent waste of oil or gas developed in the land."[17] Under this authority, in 2024 the BLM finalized a rule limiting the practice of "flaring" on public lands. If you've driven past an oil field, you've probably seen pipes coming out of the ground with a constant flame at the end—that's flaring. Oil producers deliberately ignite unwanted natural gas because gas is much more expensive to capture than liquid crude, and because accumulated gas poses the risk of explosion. Flaring gas is preferable to venting it because combustion converts methane into carbon dioxide, which is the less potent greenhouse gas of the two.

The amount of methane flared from oil fields can be truly staggering, especially when the price of natural gas is relatively low (as it was throughout the 2010s) and the economic incentive to capture it is minimal. You can imagine why an oil company would flare gas if the alternative is to build a system of gas pipelines at the cost of millions of dollars. Still, many Americans quite reasonably regard the practice as wasteful and inefficient.

Closure and Reclamation

Land management agencies must continue to manage public lands even when mines close down and oil and gas wells run dry. These agencies don't want to clean up messes made by other people, but historically the problem of rehabilitating land after mine or well closure has been severe. There are many reasons for this. First, the public lands are littered with abandoned mines and wells that predate the agencies' modern management. The BLM alone estimates that there may be half a million abandoned mines on the land it manages. Second, the very structure of extractive industry is such that the incentive to clean up is small. From a business perspective, there is no money to be made in cleaning up a mine or well site. Even an extraordinarily good cleanup job is unlikely to provide any commercial advantage. It may well go entirely unnoticed.

Finally, it's very difficult, as a matter of public policy, to establish effective rules for closure and reclamation. A typical mine's productive life may span 50 years. Given our rapidly changing knowledge of the world, it is a serious challenge to specify cleanup or reclamation rules that will be considered adequate when they are carried out. (Fifty years *ago*, for example, we were unaware of the degree to which contaminated mine runoff can permeate aquifers.) And enforcing rules about cleanup is just as hard as writing them, if not harder. When mines or wells close, it's often because the value of a particular commodity has dropped or a mine has been exhausted. In short, mine closure often occurs at the end of the business life cycle, when revenue is scarce. Regulators try to compel firms to set aside money for reclamation ahead of time, but this too is a thorny task. A common policy tool is the *reclamation bond*. A bond is a bit like an insurance policy in that a financial entity agrees to pay a certain amount if the bond purchaser defaults on their cleanup obligations. But again, it's very hard for regulators to know how much money to require for a bond that might not be claimed for a half century—and as you can imagine, the industry fiercely opposes increases to bond requirements. The result is that reclamation bonds chronically undersupply the funds necessary to reclaim derelict mines and wells.

This problem is on full display in the decline of the coal industry. In the late 2010s, nearly every major American coal producer went through bankruptcy proceedings. During those proceedings, many producers noted their likely inability to satisfy their cleanup obligations. They argued that, if those obligations were enforced, they would have no choice but to go out of business entirely—and that the only way to preserve a revenue stream for future cleanup was to diminish their environmental liability. Bankruptcy courts and regulators often felt they had little choice but to relieve firms from some amount of liability on the theory that half a loaf is better than no loaf at all. Over the years, Congress has sometimes supplied federal funds for cleanup, but this "solution" comes at taxpayer expense, of course.

BOX 13.5 When is it Time to Close an Oil or Gas Well?

It might seem obvious that oil and gas wells should be capped and sealed when they stop producing. But that's not how wells work: they don't just one day stop producing. Instead, production declines gradually. Even one of the first oil wells ever drilled in the U.S. is still producing a trickle of oil. (You can visit it in Titusville, Pennsylvania.) The question of when an under-performing well should be fully decommissioned is surprisingly tricky. In general, public policy dictates that decommissioning should occur when a well is unprofitable. But profitability depends on the price of oil, which changes all the time. Even wells that have been unprofitable for years might become profitable again during a spike in oil prices. This is yet another reason why well owners often resist well closure. It's an expensive process to begin with, and given that it costs nothing to wait and watch for price changes, you can imagine why many producers simply choose to delay proper well closure.

Renewable Energy and the Public Lands

Increasingly, energy-related interest in the public lands comes not from fossil fuel companies but from renewable energy producers. Many millions of acres of public land are well-suited for solar and wind energy. Many more are needed for the infrastructure associated with renewables, like the transmission lines that transport energy from desolate spots to crowded cities. Still more public land acreage contains minerals, like lithium or cobalt, which are essential for batteries or wind turbines. Just a few decades ago, public land managers had little experience with wind or solar power, and their administrative processes reflected this inexperience. Now, the processes for renewable-related public land use are well-established.

Wind and Solar Installations

The Desert Southwest is perfect for solar energy generation. There are vast public lands in California, Arizona, Nevada, Utah, and New Mexico that many regard as ideal for solar farms. Similarly, plenty of public land in Wyoming, Oregon, and Nevada has significant wind energy potential. FLPMA gives both the Interior and Agriculture Secretaries broad discretion to permit "systems for generation, transmission, and distribution of electric energy"[18] on public lands. Somewhat confusingly, this authority derives from FLPMA's flexible "right of way" provisions. A right-of-way is an "easement, lease, permit, or license to occupy, use, or traverse public lands"[19] for purposes allowed by the statute. The law requires applicants to submit their plans for review well in advance and to pay fair market value for the use of federal land. If the agency chooses to issue a right-of-way, it is obligated to include provisions that "minimize damage to scenic and esthetic values and fish and wildlife habitat and otherwise protect the environment."[20]

While the legal basis for siting solar and wind energy on public land is sound, there have nonetheless been challenges. The most obvious difficulties arise when an energy company wants to use a parcel of public land that poses a particular resource conflict. For example, land useful for a solar installation might also be valuable habitat for an endangered species like the desert tortoise. Conflicts like this can trigger extensive agency processes under NEPA and the ESA, leading to years of delay. To speed things along, the Interior Department has tried to classify its lands ahead of time, grading them according to their suitability for renewable energy use. This classification process hasn't been entirely smooth either, not only because of disputes over specific judgments made by the agencies but also because those agencies sometimes fail to understand or predict what makes land desirable to an energy developer. Commercial entities have a much better understanding of a project's economics than the land manager does. As an example, when the BLM held its first competitive lease sale for a solar energy zone—in the San Luis Valley of Colorado—there was not a single bidder. The BLM had, it seems, significantly misjudged the industry's interest in that zone.

But even while the kinks are being worked out, plenty of renewable development is occurring on public lands. The numbers are growing rapidly. There are already many gigawatts of solar energy-generating capacity on public lands, and wind energy is not far behind.

Transmission Lines

Transmission lines are just as important to the renewable revolution as electric generating facilities. Solar and wind installations are useless until they are connected to the grid. In an important sense, transmission lines are a substitute for new power plants. Because electricity demand varies geographically according to factors like temperature and time of day, users connected to a large grid can take advantage of spare capacity from plants whose full output is not needed in their immediate locale. That's one of the primary reasons we have an interconnected power grid.

But connecting remote renewable sites to the grid has been a major challenge. For one thing, transmission lines pose a chicken-and-egg problem. Renewable developers are reluctant to build solar or wind farms without knowing that transmission lines exist to bring their power to market, but transmission developers are unwilling to build new lines until they are confident that new plants will put the lines to use. Getting legal permission to site transmission lines is also notoriously difficult. Lines cross hundreds of miles and typically require the use of governments' eminent domain authority (that is, their power to force landowners to sell specific parcels of land), which can be deeply unpopular. Furthermore, if a line's benefits flow mostly to residents of another state, the state hosting the line may be disinclined to permit it. For all these reasons, there is a well-publicized shortage of transmission capacity.

To many observers, public lands can partially solve this problem. The federal government presently lacks the authority to site transmission lines on state or private lands, but on the public lands it can do what it wants. Federal agencies can straightforwardly use the FLPMA right-of-way process described above to put lines where they are needed most. Of course, issues still arise in connection with endangered species, scenic protection, and other resource conflicts. But many deem the federal siting process superior to the states', especially when a single federal approval can provide more linear miles for transmission than dozens of negotiations with various state and private landowners.

Critical Minerals

Shifting away from fossil fuels means shifting toward a different mix of minerals, most of which are locatable under the Mining Act. Batteries for hybrid and electric vehicles require nickel, lithium, and cobalt; wind turbines need rare earth elements like neodymium and dysprosium; and so on. Many of these minerals are primarily sourced from other countries, leaving U.S. supply chains vulnerable. Pressure is rising to produce these minerals on federal public lands, but there are counter-pressures, too. As we've seen, mining always raises environmental concerns, and some desirable deposits lie underneath Indigenous sacred sites. On the other hand, many argue that overseas mining is less environmentally protective and more likely to entail human rights violations than domestic mining. One thing seems certain: in the coming years, federal land will be thoroughly scrutinized for new mineral deposits, and the Mining Act's location system and all its idiosyncrasies will once again be in the public eye.

Conclusion

For as long as the federal government has existed, its lands have been exploited for their mineral and energy resources. The General Mining Act is a creature of its time: whether due to necessity or nepotism, Congress by the 1870s had grown accustomed to giving away frontier resources to those who claimed them. But the Act has proven almost comically durable. Whereas other aspects of public land law have become thoroughly modernized, the Mining Act lives on, despite a steady parade of opposition and the Act's poor fit with modern mining. Critics of the Act point to the absurdity of giving away valuable minerals for free—several billion dollars' worth each year—at the instigation of the mining industry. Every few years, several members of Congress mount an assault on the Mining Act, armed with tales of the Act's abuse and backed by prominent op-ed pieces. And yet the Act survives, thanks to industry support, to intransigence among the Western states, and perhaps most of all to the general apathy of the public toward this obscure corner of public policy.

Although the leasing system has its warts, too, it does return money to federal and state coffers, and for that reason alone it's likely that fossil fuels will remain in production on public lands for quite some time. But renewable energy is rising rapidly, and where public lands were once assessed for their coal, oil, and gas potential, today they are just as likely to be prized for their suitability for wind, solar, or transmission. No doubt the future will elevate interest in minerals and energy-related land uses that we haven't yet begun to consider. The mineral abundance of the federal public lands guarantees their place in such conversations long into the future.

Study Questions

1 What are the fundamental differences between the mineral location system under the 1872 Act and the mineral leasing system under the 1920 Act?
2 Why do critics argue that the General Mining Act of 1872 is an outdated law that unfairly subsidizes the hardrock mining industry?
3 What constrains the regulatory power of the federal land management agencies when overseeing mineral development?

Resources for Further Study

Bakken, Gordon M. *The Mining Law of 1872: Past, Politics, and Prospects* (Albuquerque, NM: University of New Mexico Press, 2008).

Brennan, Bailey, Jim Murphy, and David DeGennaro. *A Guide to Responsible Development of Wind and Solar Resources on Public Lands and Waters* (Washington, D.C.: National Wildlife Federation, 2023).

Greene, Mary. "Reforming Oil and Gas Leasing on Public Lands." *Vermont Law Review* 45 (2021): 575–590.

Leshy, John D. *The Mining Law: A Study in Perpetual Motion* (New York, NY: Routledge, 1987).

Squillace, Mark. "The Minerals Challenge for Renewable Energy." *Environmental Law Reporter* 54, no. 1 (Jan. 2024): 10058–10079.

Notes

1 Throughout this chapter, we use the word "minerals" to include even liquids (like oil) and gases (like natural gas). Historically, "mineral rights" encompass all these valuable substances, even though now we typically think of "minerals" as solid substances.
2 John Leshy, *Our Common Ground: A History of America's Public Lands* (New Haven, CT: Yale University Press, 2021), 553.
3 John Lancaster, "Mining Act Abuse: Rent-Free Living on Public Land," *The Washington Post* (Aug. 30, 1990).
4 U.S. General Accounting Office, "Hardrock Mining on Federal Lands" (Sept. 6, 1990).
5 If you've never seen something like this, peek at an image of a pit mine on the internet.
6 Congressional Research Service, "The U.S. Mining Industry and the *Rosemont* Decision" (CRS Report R48166, Aug. 26, 2024).
7 Statement of Steve Feldgus, U.S. Dept. of the Interior, before the House Natural Resources Committee, Subcommittee on Energy and Mineral Resources, "Reforming the Mining Law of 1872" (May 12, 2022).

8 Because federal law can't possibly be straightforward and consistent, there are some lands upon which hardrock minerals are leasable rather than locatable. See 43 CFR Part 3500 ff.

9 An oil and gas lease is quite different than, say, a lease for a residential apartment, but they are similar in that the lessor retains permanent ownership of the property.

10 U.S. Geological Survey, *Assessment of Undiscovered Oil and Gas Resources of the Powder River Basin Province of Wyoming and Montana—2006 Update* (Washington, D.C.: U.S. Dept. of Interior, 2006).

11 This is called the "lease by application" process. Federal law also allows BLM to lease tracts of its own choosing, but the BLM has relied exclusively on "lease by application" since 1990. See Mark Squillace, "The Tragic Story of the Federal Coal Leasing Program," *Natural Resources & Environment* 27 (2013): 29, 35.

12 Office of the Inspector General, U.S. Dept. of the Interior, "Coal Management Program" (Interior Dept. Report No. CR-EV-BLM-0001–2012, 2013); Tom Sanzillo, "The Great Giveaway: An Analysis of the United States' Long-term Trend of Selling Federally Owned Coal for Less than Fair Market Value" (Institute for Energy Economics and Financial Analysis, 2012); Taxpayers for Common Sense, "Federal Coal Leasing: Fair Market Value and a Fair Return for the American Taxpayer" (2013).

13 U.S. Government Accountability Office, "Coal Leasing: BLM Could Enhance Appraisal Process, More Explicitly Consider Coal Exports, and Provide More Public Information" (GAO Report GAO-14–140, 2013).

14 Bureau of Land Management, "Rock Springs Field Office: Draft Resource Management Plan Revision and Draft Environmental Impact Statement" (May 2023).

15 Bureau of Land Management, "Rock Springs Field Office: Draft Resource Management Plan Revision and Draft Environmental Impact Statement".

16 These regulations can be found in 36 CFR § 228 subpart A.

17 Mineral Leasing Act § 225.

18 43 U.S.C. § 1761(a)(4).

19 43 U.S.C. § 1702(f).

20 43 U.S.C. § 1765(a)(ii).

Afterword

We began this book with a simple premise: the public lands are a choice, an ongoing political project shaped by geology, ecology, and—most profoundly—the collective and conflicting desires of the American people.

Now that you have journeyed from the fraught acquisition of the public domain in the nineteenth century to the complex regulatory environment of the twenty-first, we hope the stakes and the magnitude of that choice are more clear. You have seen how the jumbled land laws described in the early chapters hardened into the distinct systems we know today: the National Parks, with their dual mandate of preservation and enjoyment; the National Forests and their utilitarian roots; the Wildlife Refuges and their specific biological imperatives; and the vast, multiple-use lands of the BLM. Each of these systems meets today's challenges in different ways, working within the legal constraints forged over decades by Congress, the courts, and the executive branch. Those challenges are profound, and the tools the law provides are not always fit for purpose.

In the Preface, we noted that we have written this book during a time of extraordinary dynamism. The political pendulum, which once swung in slow, decadal arcs, does not seem pendular at all. In just the last few years, we have watched as fundamental regulations were finalized, implemented, and then rescinded or rewritten, sometimes within the span of a single presidential term. We have seen the staffing levels of agencies like the Forest Service fall precipitously, reshaping the capacity of the federal government to manage the lands it holds in trust.

This volatility validates the warning we offered at the outset: the laws and policies governing public lands are not static. Laws like the General Mining Act of 1872 or the Wilderness Act of 1964 provide a legal foundation, but the structure built upon it is constantly being renovated, expanded, or stripped down. A national monument designated today may be reduced tomorrow; a prohibition on mining in a sensitive watershed may be reversed by a new administration or a rider on an appropriations bill.

Yet amidst this turbulence, the land remains, and it continues to demand our attention. The peaks of the High Sierras, the coal seams of the Powder River Basin, and the migration corridors of the pronghorn still shape our political ecology, particularly in the West. Whether you are a student contemplating a career in land management, a voter considering a ballot initiative, or simply a visitor hiking a trail in a National Forest, you are a participant in the story of the public lands. Their future is not written in stone; it is written in the laws we pass, the leaders we elect, and the values we choose to uphold.

While this book is not a manual about how citizens can shape the direction of public lands, it provides the history and context necessary to engage in the public lands debate. You now understand why a checkerboard map pattern complicates access in

DOI: 10.4324/9781003392514-14

Wyoming; you recognize the legal difference between a locatable and a leasable mineral; you can discern the distinct statutory missions that might lead a park superintendent to make a different decision than a regional forester. The land agencies are subject to a mix of demands, but there are many opportunities to be involved in their decisions. The challenges we have explored in these pages, from the existential threat of climate change to the localized conflicts over recreation and resource extraction, will not be solved by a single statute or a court ruling. They will be managed, day by day, by the public servants, scientists, and citizens who engage with the democratic processes we have described.

Index

Note: **Bold** page numbers refer to **tables**; *italic* page numbers refer to *figures*.

For Product Safety Concerns and Information please contact our EU representative GPSR@taylorandfrancis.com Taylor & Francis Verlag GmbH, Kaufingerstraße 24, 80331 München, Germany